AF885465

INTERIOR OPTIMIZATION METHODS WITH ELECTRONICS APPLICATIONS EXAMPLES

VOLUME I

INTERIOR OPTIMIZATION METHODS WITH ELECTRONICS APPLICATIONS EXAMPLES

VOLUME I

Francisco Casesnoves

ELIVA PRESS

ISBN: 978-99949-8-689-7

Cover Design: Eliva Press
Cover Image: Ingimage
Email: info@elivapress.com
Website: www.elivapress.com

This Eliva Press imprint is published by the registered company Eliva Press Global Ltd. part of Eliva Press S.R.L. Publishing Group

The registered company address is: Pope Hennessy Street Level 2, Hennessy Tower Port Louis, Mauritius
Eliva Press S.R.L. Publishing Group address is: Bulevardul Moscova 21, Chisinau, Moldova, Europe

INTERIOR OPTIMIZATION MATHEMATICAL METHODS WITH ELECTRONICS APPLICATIONS EXAMPLES [VOLUME I]

Francisco Casesnoves

PREFACE

This First Volume Interior Optimization Series book presents the enhanced/improved fundamental and initial articles of Interior Optimization Methods publication series with direct applications on Electronics Physics examples. In plain language, Interior Optimization is a general numerical computational-graphical method based on graphical variables separation techniques. That process is applied in subsequent 3D imaging stages, and could be also 2D. That is, the analytical partial differential equations method basis is used to make a graphical-optimal numerical determination-selection. This is made through 3D/2D imaging-stages in order to select the best/desired values of any several variables algorithm. The mathematical and computational fundamentals of the method are shown. Initially, these techniques were designed for superconductor modelling optimization, specifically BCS Isotope Effect optimization, the Molecular Effect Model (MEM), and Superconducting Multifunctional Transmission Lines Project, namely, Chapter IV. This Chapter IV includes one Appendix with recent results for Molecular Effect Model. The primary published contributions are revised and got better. Every chapter comes from a previous published article, complemented, and reviewed. Further explanations of computational software methods in Matlab® and GNU-Octave® (Chapter II), are included. An exception is Chapter IV, which is included jumping the chronological articles publications series. The reason is to set the importance of the engineering application for Interior Optimization methods for the Superconductor Multifunctional Transmission Line. All these optimization approaches can be utilized for any science branch that could demand this necessary type of mathematical-computational methods. In other words, the reader can obtain from the electronics applications sufficient learning to use Interior Optimization in other science field.

To the Pursuit of the Truth for Itself,

Tallinn, 22th August 2022, 11 January 2023,

Dr Francisco Casesnoves PhD

CONTENTS

CHAPTER I

MATHEMATICAL BASIS AND FIRST APPLICATIONS

ABSTRACT

The nonlinear multiobjective optimization problem constitutes a difficult task to find/determine a global minimum, approximated-global minimum, or a convenient local minimum whith/without constraints. Nonlinear systems of equations principles set the base for further development of Interior Optimization and Interior-Graphical Optimization [Casesnoves, 2016-7]. Mathematical proofs with mathematical definitions, algorithms, and nonlinear optimization equations are presented. From Graphical Optimization 3D optimization stages, it is proven that solution of nonlinear systems of equations is not unique in general. Software-engineering and computational simulations are shown with electronics superconductors and electronics physics applications. Extensions to similar applications for materials-tribology models and Biomedical Tribology are explained.

Keywords : Interior Optimization Methods, Graphical Optimization, Systems of Nonlinear Equations, Electronics

1. INTRODUCTION

In optimization with objective functions based on nonlinear systems of equations [3,4,6,12-15,18,30], or multiobjective-parameters constraints and other conditions, the task is rather difficult. Specifically this occurs in Physics and Engineering nonlinear optimization problems. Large-scale amount of empirical data for algorithms/models implementation have to be adapted/fitted to models that involve a high number of equations, nonlinear many of them [12-15], constraints added. Stochastic Optimization methods are used to save time and computational effort when the amount of data is large. In these cases, it is not necessary to work with all the numerical volume available.

Approximations and final refinements for mathematical models are usually performed with inverse methods [2-4,12-15]. Statistical-estimations/Approximations and errors testing are obtained from all alternative formulation available. In previous contributions, [2-4,19-27], Graphical Optimization was presented and applied for engineering mathematical modelling, materials erosion models [19], and physics of deformable solids dynamics.

However, Graphical Optimization involves the objective function visualization in 2D and 3D. This implies that multiparameter formulas initially cannot be easily set for the 3D Graphical Optimization, unless a choice of 2 or 3 variables is done among several ones. From this apparent difficulty, a mathematical optimization and software engineering method for an objective function of high number of variables arose and is presented. It is defined as Interior-Graphical Optimization Method [Casesnoves, 2018]. This observed fact implied also that some complementary approaches to the theory of Nonlinear Equations systems could be done.

Furthermore, in this line, it is proven for optimization that the solution for multiobjective optimization problems is not unique in general. This occurs in nonlinear systems equations [22]. Mathematical development and algorithms are graphically-shown to demonstrate the former assertion. The method was intended to be proven and set, at first, on simple equations at Electronics Physics field, namely BCS algorithms [22, 31-37].

Therefore, direct applications in electronics field for superconductors isotope-effect and critical temperature are presented [2-4,12-15,19-27].
In summary, this continued-study shows an advanced contribution for the Interior Nonlinear Optimization Method previously presented in [3]. Mathematical and computational base is explained/proven at the Superconductors Type I area. The scheme of the method is shown, with a sharp proof in applications at this electronics field.

2. MATHEMATICAL METHODS AND ALGORITHMS

Linear systems are included in this study as a nonlinear ones with exclusively unitarian power and without nonlinear operations among parameters. Consider the multiobjective problem with constraints,

$$
\begin{aligned}
&\text{minimize,}\\
&f(\vec{x}) = f\left(f_1(x_1), f_2(x_2), \ldots\ldots f_N(x_N)\right) \quad ,\\
&\text{subject to,}\\
&a_1 \le x_1 \le b_1\\
&a_2 \le x_1 \le b_2\\
&\cdot\\
&\cdot\\
&\cdot\\
&a_N \le x_N \le b_N \;;
\end{aligned}
$$

(1)

where,

$f(x)$: Function of several functions with several variables. X is a vector whose components are several (N) functions variables. For simplification, it is chosen that every function $f_i(x_i)$ is function of exclusively one variable x_i .

X_i : Every variable of function $f(x)$.

N : Number of components of vector X, which is the number of variables.

a_i : Every [i] minimum value of the N constraints.

b_i : Every [i] maximum value of the N constraints.

This (1) is an equation of several variables that has not, in general, unique solution. If the objective function becomes an [i] number of nonlinear functions [11-15]. Hence,

$$
\begin{aligned}
&\text{for } i=1,\ldots N,\\
&\text{minimize,}\\
&f(\bar{x})=f(x_{i1},x_{i2},\ldots\ldots x_{iN}) \quad ,\\
&\text{subject to,}\\
&a_{i1} \le x_1 \le b_{i1}\\
&a_{i2} \le x_1 \le b_{i2}\\
&.\\
&.\\
&.\\
&a_{iN} \le x_{iN} \le b_{iN} \;;
\end{aligned}
$$

(2)

where,

f (x) : Function of several functions with several variables. X is a vector whose components are several (N) functions variables. For simplification, it is chosen that every function f_i (x_i) is function of exclusively one variable x_i .
X_i : Every variable of function f (x).
N : Number of components of vector X, which is the number of variables.
a_{iN} : Every [i] minimum value of the N constraints for function f_i .
b_{iN} : Every [i] maximum value of the N constraints for function f_i .

Therefore, it is a system of nonlinear equations functions with several variables, [11-15,26], that in general case do not have to be equal in number to the parameters. If there are terms with operations among parameters, and roots, the unique solution is even more difficult to be find out. Optimization algorithms and software can find usually a local minimum, semi-global minimum, or a local minimum according to constraints [11].

A classical method is Monte-Carlo programs, such as GEANT type [11-15] used in IMRT-IMPT radiotherapy, although there are options for Monte-Carlo and Quasi-Monte-Carlo software available. Jacobian and classical Newton-Raphson methods are, for example, used to find an approximate solution. Number of methods to find solution/approximate solutions for a system of nonlinear equations are extent, and it is not the specific focus of this contribution. Monte-Carlo method can be considered methods to find solution/approximate solutions for a system of nonlinear equations are extent [26], and it is not the specific focus of this contribution. Monte-Carlo method can be considered good, because the program usually, taking continuous random values, search for an optimal solution of variables, no matter how many they are. Genetic Algorithms constitutes an stochastic method that has proven be very useful and precise nowadays.

Given this algorithm (1-2), the Graphical Optimization Method can be applied starting for grouping variables step by step. The technique is an extrapolation of the classical partial equations method of separation of variables. That is,

If

$$f(\vec{x}) = f(x_1, x_2, \ldots\ldots x_N)$$

(3)

The variables can be grouped in such a way, for instance,

if we find, X_1 multiplied by X_2 ,

it is defined, $X_{1s} = X_1 \bullet X_2$;

(4)

Or just the same parameters-grouping for sums, exponentials, etc. Thus, the technique begins with a 3D graph of two selected parameters, at Z axis the objective function [in 13, the beginning of method, ch1]. The optimal/desired values are fixed at that graph. Those fixed values are taken for the subsequent 3D graph that involves the following two parameters, and so on.

Definition I.-Interior-Graphical Optimization Method, [Casesnoves, 2018] is a type of Nonlinear Optimization that combines separation of variables method with stages of Graphical Optimization [Casesnoves, 2016], [Interior Graphical-Optimization Methods were created by Francisco Casesnoves on 3rd November 2018, while he was preparing his Doctoral Thesis defence. First implementation of algorithms and computational-verification of simulations were carried out in the morning of April 1st, 2020].

What is done in mathematical concepts, is to apply the partial differential equations classical separation of variables method. With separation of variables, it is possible to optimize all the parameters in subsequent stages of 3D graphical optimization plots. At every plot, it is selected the desirable local, global, or semi-local minimum for the variable of convenience [ch1-4].

In such a way that the initial Graphical Optimization begins with 3 variables, namely, first, second and objective function. An optimal value is determined. The following stage one parameter with two variables is decomposed, and it is taken the optimal values, and so on.

Therefore, it is easy to prove, through the multi-selection of local and approximate minima at Graphical Optimization sets, that in general the multiobjective solution for a system of nonlinear equations is not unique. [30].

In other words, it is demonstrated from 3D visualization and separation of variables, this classical mathematical assertion. Number of graphs required depend on the number of variables of the nonlinear system of equations. The determination of optimal values not necessarily have to be minima. Optimal values taken depend of the engineering requirements for laboratory optimization, devices manufacturing, experimental plan, or similar tasks. The advantage is that the software engineering gives a fast tool to find necessary data immediately with short time consuming [ch1, 19-27].

3. ELECTRONICS APPLICATIONS ON SUPERCONDUCTORS

In this section a brief summary of Superconductivity towards the Transition Equation and the Isotope Effect in BCS Theory is presented. The Superconductivity Theory framework is rather extent and with large mathematical background. It involves Quantum Theory, Molecular Chemistry, Materials Physics and other science and engineering branches [31-37].

All these areas converge together in Superconductors theory and applications. Which is useful to prove Interior Optimization

The method are the simple equations of Transition-Temperature-Critical and Isotope Effect [ch1-4, 2-4, 31-37]. The classical-simple Critical Temperature Equation for superconductors reads,

The classical-simple Critical Temperature Equation for superconductors reads,

$$[M_i]^{\alpha} T_{Ci} - K \cong 0 ;$$
$$\text{for } i = 1, \ldots .n ;$$

(5)

where,

K : Constant parameter. Range specific for every element.
M_i : Atomic Mass (AMU) of any element isotope if (n) isotopes.
α : Constant parameter. Range specific for every element.
T_C : Critical temperature (K (usually), or C for High-Temperature Superconductors). Range specific for every element.
i : Corresponding isotope for every element.

From the stage when the solid reaches critical temperature, the superconductivity effect begins. T_C varies for every isotope element within a defined mass element, although not excessively in magnitude.

This equation, although simple, remains useful today. It was taken to illustrate Interior Optimization Method in simple way. The evolution of this equation, for example, [19], involves extent mathematical background and mathematical-physics theory. It is sufficient to prove the Interior Optimization Method with this classical equation.

Taking decimal or Neperian logarithms the equation (5) can be linearized for setting graphics related to different isotopes with corresponding atomic proper mass. To set the simplest nonlinear system of equations, the algorithm selected with 100 equations reads,

$$
\begin{aligned}
&\text{minimize} \\
&\left\| [M_i]^{\alpha}\, T_C - K_i \right\|_1 \\
&\text{for } i = 1,\ldots..n \\
&\text{subject to} \quad , \\
&a \le M_i \le a_1 \,; \\
&b \le T_C \le b_1 \,; \\
&c \le K_i \le c_1 \,; \\
&d \le \alpha \le d_1 \,;
\end{aligned}
$$

(6)

where

K : Constant parameter. Range specific for every element.
α : Constant parameter. Range specific for every element.
$|\cdot|_1$: Chebyshev L_1 Norm. (at algorithm power 1).
a, a_1 : Constraint range specified at Table 1.
b, b_1 : Constraint range specified at Table 1.
c, c_1 : K optimization parameter range for program.
d, d_1 : α constant range for program.

where i is the corresponding number within the range of simulations. For example, in range of simulations for Hg, M [ch1] is,

$M \in [198.0\ ,\ 203.4]$

(7)

In the same way, range of constraints can be seen in axes ranges of Figs 1-2. That is, 100 divisions in M range, (7)

that comprise the variation of atomic mass for all Hg isotopes within. Therefore, with equation (6), a nonlinear system of equations is set. The variables are to get optimized. The method is Interior-Graphical Optimization, that will be proven accurate and useful for (6). It was taken Hg superconductivity values from literature. Extrapolation to large and complicated Electronic Physics equations, Electronics, or any type of system can be guessed from this application.

4. SOFTWARE ENGINEERING AND MATRIX ALGEBRA ALGORITHMS

The software developed for this contribution had some easy parts and not a few programming difficulties to adapt the programs on Chebyshev L_1 norm for objective function (6) . Additional complications were the algebra changes in the subroutines to get functional the program for surface Graphical Optimization visualization. This requires mathematical changes-background.

The software was developed in Matlab and is almost equivalent for Freemat also. These numerical data are all for this contribution. Range for surfactal simulations Delanuay tiles was 100. In previous contributions, higher values were taken for high-accuracy.

However, optimization programs were designed for double precision, since were based on engineering software of extent number of previous publications [2-4,19-27].
To check complementarily the numerical results, both Fortran and F# subroutines in optimization were used. While Fortran proved to match the Graphical Optimization results, F# showed restrictions and obviously, limits that discarded this type of language for accurate results in nonlinear optimization. None of these numerical data were included in publication. F# actually cannot be considered a rigorous scientific or engineering software programming language, rather a business-commercial management language.

5. COMPUTATIONAL GRAPHICAL-INTERIOR OPTIMIZATION RESULTS

The results are shown in Tables and Graphics. Numerical results matches values in the literature for natural Hg. However, alpha value is lower than the classical magnitude of 0.5.

Residuals can be considered acceptable, although not of high-precision accuracy. The time consuming for designing program is extremely short. The time for selecting optimal values is also very short with the Matlab graphical cursor. In engineering practice, this is a significant advantage that gives many options for fast on-site superconductivity data management at laboratory. Results confirm useful utility of equation (5).

NUMERICAL INTERIOR OPTIMIZATION RESULTS I		
Hg ISOTOPE TYPE (BY ATOMIC MASS)	OPTIMAL ALPHA	FIXED T_C (Kelvin)
200.7 (NATURAL)	0.1455	4.154
199.9	0.1455	4.154
202.0	0.1455	4.154
203.2	0.1455	4.154

Table 1.-The first stage of Interior Optimization. In X and Y axes, are set the constant K and the parameter A that comprises Mα. That is, A= M^{α} in Fig. 1. The values are taken almost from all references related to Hg critical temperature for superconductors. T_C value is fixed for natural Hg in simulations and at Z axis it is deployed the Objective Function without taking absolute value. T_C for natural Hg is fixed because it is searched optimal values for all isotopes conditioned to the Tc of the natural Hg. The results for optimal K are in Table 2. These values were taken with the cursor in the most easy and simple way. It took about 10-20 minutes. Programs construction were performed in 2 hours based on previous software [13] with matrix algebra modifications.

NUMERICAL INTERIOR OPTIMIZATION RESULTS II		
Hg ISOTOPE TYPE (BY ATOMIC MASS)	K OPTIMAL	OBJECTIVE FUNCTION RESIDUAL (Chebyshev Optimization Norm)
200.7 (NATURAL)	57.06	0.5891
199.9	57.06	0.5891
202.0	57.06	0.5891
203.2	57.06	0.5891

Table 2.- The second stage of Interior Optimization. In X and Y axes, are set the constant α and the parameter M. That is, previously was optimized $A= M^{\alpha}$. The values are taken almost from all references related to Hg critical temperature for superconductors. T_C value is fixed for natural Hg and at Z axis it is deployed the Objective Function without taking absolute value.

The results for optimal α and residuals are in Tables 1 and 2. These values were taken with the cursor in the most easy and simple way. It took about 10 minutes.

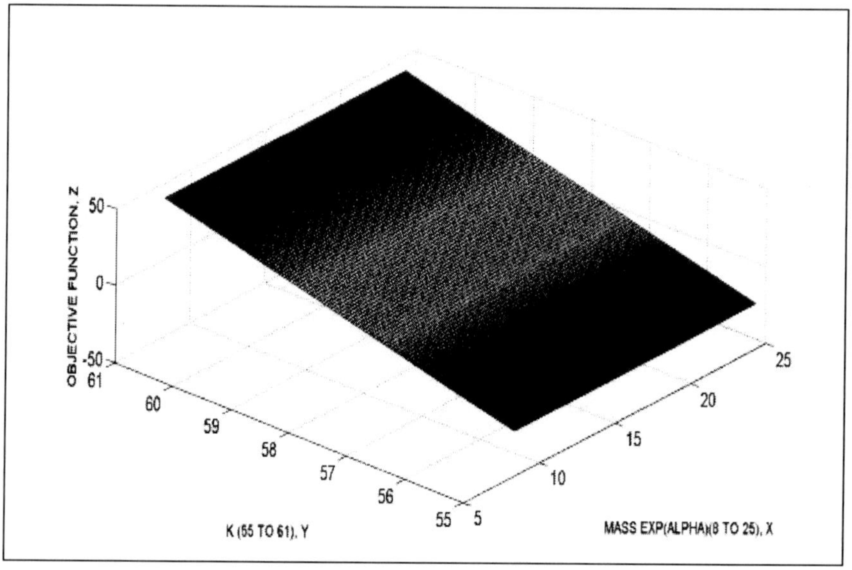

Fig. 1.-First stage of Hg Interior-Graphical optimization. It was determined the optimal K value and the optimal A= M^{α} value. Information and tools that can be used/obtained with imaging-surfactal subroutines which are detailed in [ch1-4].

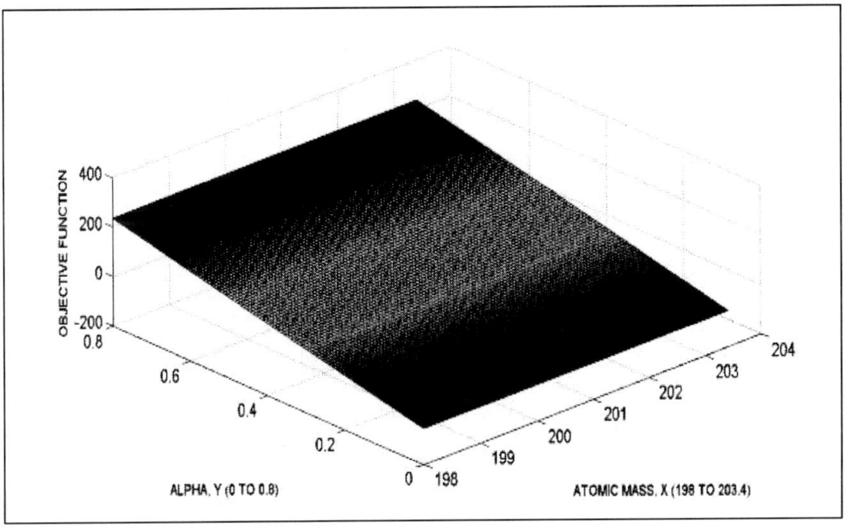

Fig. 2.- Second stage of Hg Interior-Graphical Optimization. It was determined the optimal M value and the optimal α value. Information and tools that can be used/obtained with imaging-surfactal subroutines are detailed in [ch1-4].

6. DISCUSSION AND CONCLUSIONS

The Interior Optimization Method base with algorithms and mathematical definitions was explained and presented formally. The nonlinear systems of equations linked to the theoretical algorithmic-construction was set. Its development with nonlinear optimization combined with graphical optimization [Casesnoves, 2016-2017] method constitute the theoretical fundamentals of the article.

As a demonstration of the mathematical and practical engineering applications, electronics in superconductors equations were presented. Computational electronics simulations with 3D graphical optimization images were made correctly. Numerical results were explained in tables and schemes.

Numerical results match the usual values published in literature for Hg superconductivity effect. In alpha value determination, the magnitude differs from the common 0.5 value set classically. Not always, however, 0.5 alpha value is calculated in all publications of the literature studied in [ch1-4].

In summary, original method of Interior Optimization was developed with examples practical electronics applications examples in superconductor BCS electronics field. The non-existence, in general, of uniqueness of solutions for a system of nonlinear equations from Graphical Optimization Method was demonstrated. Applications in other areas of Electronic Physics, Physics and Engineering emerge from this Interior Optimization Method.

Additional illustration applications examples in other research fields were presented/proven [2-4]. In Tribology and Biomedical Tribology the usage of this method is similar with the Graphical Optimization technique that was published in previous contributions [19-27]. For example, as it was presented for medical devices implants optimization [27].

Original Article Reference

Francisco Casesnoves , "Interior Optimization Methods with Electronics Applications", International Journal of Scientific Research in Science, Engineering and Technology (IJSRSET), Online ISSN : 2394-4099, Print ISSN : 2395-1990, Volume 7 Issue 3, pp. 428-436, May-June 2020. Available at DOI : https://doi.org/10.32628/IJSRSET207377 Journal URL : http://ijsrset.com/IJSRSET207377

CHAPTER II

ADVANCED INTERIOR OPTIMIZATION METHODS WITH ELECTRONICS APPLICATIONS

ABSTRACT

This Chapter deals with the beginning of further development of Interior Mathematical Method. In a previous Chapter I, the mathematical-computational base of Interior Optimization Method was demonstrated. Electronics applications were performed with numerical optimization data and graphical proofs. In this evolved-improved paper a series of electronics applications of Interior Optimization in superconductors BCS algorithms/theory are shown. In addition, mathematical developments of Interior Optimization Methods related to systems of Nonlinear Equations are proven. The nonlinear multiobjective optimization problem constitutes a difficult task to find/determine a global minimum, approximated-global minimum, or a convenient local minimum whith/without constraints. Nonlinear systems of equations principles set the base in the previous article for further development of Interior Optimization and Interior-Graphical Optimization [Casesnoves, 2016-7].

From Graphical Optimization 3D optimization stages, the demonstration that solution of nonlinear systems of equations is not unique in general emerges. Software-engineering and computational simulations are shown with electronics superconductors [several elements, Type 1 superconductors] and electronics physics applications. Extensions to similar applications for materials-tribology models and Biomedical Tribology are explained. Engineering software with GNU-Octave® is included in imaging processing charts.

Keywords : Interior Optimization Methods, Graphical Optimization, Systems of Nonlinear Equations, Electronics Superconductors.

1. INTRODUCTION

In optimization with objective functions based on nonlinear systems of equations [30], or multiobjective-parameters constraints and other conditions, the task is rather difficult. Specifically this occurs in Physics and Engineering nonlinear optimization problems. Large-scale amount of empirical data for algorithms/models implementation have to be adapted/fitted to models that involve a high number of equations, nonlinear many of them, with constraints added.

Stochastic Optimization methods are used to save time and computational effort when the amount of data is large. In these cases, it is not necessary to work with all the numerical volume available.

Approximations and final refinements for mathematical models are usually performed with inverse methods [19-27]. Statistical-Estimations/Approximations and errors testing are obtained from all alternative formulation available. In previous contributions, Graphical Optimization was presented and applied for engineering mathematical modelling, materials erosion models [12], and physics of deformable solids dynamics.

However, Graphical Optimization involves the objective function visualization in 2D and 3D. This implies that multiparameter formulas initially cannot be easily set for the 3D Graphical Optimization, unless a choice of 2 or 3 variables is done among several ones.

From this apparent difficulty, a mathematical optimization and software engineering method for an objective function of high number of variables arose and is presented. It is defined as Interior-Graphical Optimization Method. This observed fact implied also that some complementary approaches to the theory of Nonlinear Equations systems could be done.

Furthermore, in this line, it is proven for optimization that the solution for multiobjective optimization problems is not unique in general [ch1]. This occurs in nonlinear systems equations. Mathematical development and algorithms are graphically-shown to demonstrate the former assertion [Appendix I]. The method was intended to be proven and set, at first, on simple equations at Electronics

Physics field, namely BCS algorithms. Therefore, direct applications in electronics field for superconductors isotope-effect and critical temperature are presented.

In summary, this continued-study shows an advanced contribution for the Interior Nonlinear Optimization Method previously presented in [3]. Mathematical and computational base is explained/proven at the Superconductors Type I area. The scheme of the method is shown, with a sharp proof in applications at this electronics field. A clear demonstration that Nonlinear Systems do not have unique solution in general constitutes a supplementary mathematical frame of this paper [Appendix I].

2. MATHEMATICAL METHODS AND ALGORITHMS

As a reminder of Chapter I, linear systems are included in this study as a nonlinear ones with exclusively unitarian power and without nonlinear operations among parameters. Consider the multiobjective problem with constraints,

$$
\begin{aligned}
&\text{minimize,}\\
&f(\vec{x}) = f\left(f_1(x_1), f_2(x_2), \ldots\ldots f_N(x_N)\right) \quad ,\\
&\text{subject to,}\\
&a_1 \le x_1 \le b_1\\
&a_2 \le x_1 \le b_2\\
&.\\
&.\\
&.\\
&a_N \le x_N \le b_N \;;
\end{aligned}
$$

(1)

where,

f (x) : Function of several functions with several variables. X is a vector whose components are several (N) functions variables. For simplification, it is chosen that every function f_i (x_i) is function of exclusively one variable x_i .
X_i : Every variable of function f (x).
N : Number of components of vector X, which is the number of variables.
a_i : Every [i] minimum value of the N constraints.
b_i : Every [i] maximum value of the N constraints.

This is an equation of several variables that has not, in general, unique solution. If the objective function becomes an [i] number of nonlinear functions. Hence,

$$
\begin{aligned}
&\text{for } i=1,\ldots N,\\
&\text{minimize,}\\
&f(\vec{x})=f(x_{i1}, x_{i2}, \ldots\ldots x_{iN}) \quad ,\\
&\text{subject to,}\\
&a_{i1} \leq x_1 \leq b_{i1}\\
&a_{i2} \leq x_1 \leq b_{i2}\\
&.\\
&.\\
&.\\
&a_{iN} \leq x_{iN} \leq b_{iN} \;;\\
&\text{or,}\\
&\begin{pmatrix} a_{i1} \\ \\ a_{iN} \end{pmatrix} \leq \begin{pmatrix} x_{i1} \\ \\ b_{iN} \end{pmatrix} \leq \begin{pmatrix} b_{i1} \\ \\ b_{iN} \end{pmatrix};
\end{aligned}
$$

(2)

where,

$f(x)$: Function of several functions with several variables. X is a vector whose components are several (N) functions variables. For simplification, it is chosen that every function $f_i(x_i)$ is function of exclusively one variable x_i .

X_i : Every variable of function $f(x)$.

N : Number of components of vector X, which is the number of variables.

a_{iN} : Every [i] minimum value of the N constraints for function f_i .

b_{iN} : Every [i] maximum value of the N constraints for function f_i .

The notation of these equations (1-2) can be set in complete matrices form. Constraints could be also depending on parameters.

3. ELECTRONICS APPLICATIONS ON Cr and Pb SUPERCONDUCTORS

In this section a brief summary of Superconductivity towards the Transition Equation and the Isotope Effect is presented. The Superconductivity Theory framework is rather extent and with large mathematical background. It involves Quantum Theory, Molecular Chemistry, Materials Physics and other specialities. For the magnetic interaction of superconductor materials the Electromagnetic Theory is essential. All these areas converge together in Superconductors theory and applications. Which is useful to prove Interior Optimization Method are the simple equations of Transition-Temperature-Critical and Isotope Effect [31-37] . The applications of Supeconductors in Medical Technology are described extensively in literature.

Main of them are Hadrontherapy (Protontherapy and/or Carbontherapy) and Nuclear Magnetic Resonance (NMR). Protontherapy [24], is a high-precision dose delivery technique based on the advantage of Bragg Peak radiation

distribution. It constitutes an improvement of dosimetry compared to classical IMRT.

These applications are related to the area of magnetic fields interaction with superconductors. Future applications in the future are in investigation and development. The classical Critical Temperature Equation BCS for superconductors reads,

$$M^{\alpha} \cdot T_C = K \; ; \; (K{=}E \text{ sometimes in Figures})$$

(3)

where K is a constant, M Atomic Mass (AMU), and T_C is critical temperature (Kelvin). From the stage when the solid reaches critical temperature, the superconductivity effect begins. T_C varies for every isotope element within a defined mass element, although not excessively in magnitude.

This equation, although simple, remains useful. It was taken to illustrate Interior Optimization Method in simple way.

The evolution of this equation, for example, [11], involves extent mathematical background and mathematical-physics theory. It is sufficient to prove the Interior Optimization Method with this classical equation.

Taking decimal or Neperian logarithms the equation (4-5) can be linearized for setting graphics related to different isotopes with corresponding atomic proper mass. In the literature [ch1-4], it is frequent data analysis of BCS equation with linearization plots for comparisons. To set the nonlinear system of equations, the algorithm selected with 100 equations reads,

$$
\begin{aligned}
&\text{minimize} \\
&\left\| [M_i]^{\alpha}\, T_C - K_i \right\|_1 \\
&\text{for } i = 1, \ldots, n \\
&\text{subject to} \quad , \\
&a \le M_i \le a_1 ; \\
&b \le T_C \le b_1 ; \\
&c \le K_i \le c_1 ; \\
&d \le \alpha \le d_1 ;
\end{aligned}
$$

(4)

where

K : Constant parameter. Range specific for every element.
α : Constant parameter. Range specific for every element.
$|\cdot|_1$: Chebyshev L_1 Norm. (at algorithm power 1).
a, a_1 : Constraint range specified at Table 1.
b, b_1 : Constraint range specified at Table 1.
c, c_1 : K optimization parameter range for program.
d, d_1 : α constant range for program.

where i is the corresponding number within the range of simulations. For example, in range of simulations for Pb AMU is,

$M \in [204.0 , 208.0]$

(5)

In the same way, range of constraints can be seen in axes ranges of Figs 1-6. That is, 100 divisions in M range, that comprise the variation of atomic mass for all Hg isotopes within.

Therefore, with equations 1-5, a nonlinear system of equations is set. The variables are to get optimized. The method is Interior-Graphical Optimization, that will be proven accurate and useful for (6). It was taken Hg, Pb, and Cr superconductivity values from literature. Extrapolation to large and complicated Electronic Physics equations, Electronics, or any type of system can be guessed from this application.

4. SOFTWARE METHOD AND ALGORITHMS IMPLEMENTATION

The algebraic software developed for this contribution had various programming difficulties to adapt the programs on Chebyshev L_1 norm for objective function. Additional complications were the algebra changes in the subroutines to

get functional the program for surface Graphical_Optimization visualization. This requires mathematical changes-background. The software was developed in Matlab and is almost equivalent for Freemat® also. These numerical data are all for this contribution. Programming was selected to use subroutines that let rotate 3D images to obtain best visualization. With this tool it was possible to see in images the sharp minimum zone to prove the mathematical statements. Freemat also offers this possibility in 3D surfactal implementation. Range for surfactal simulations Delanuay tiles was 100. In previous contributions, higher values were taken for high-accuracy. However, optimization programs were designed for double precision, since were based on engineering software of extent number of previous publications [2-4].

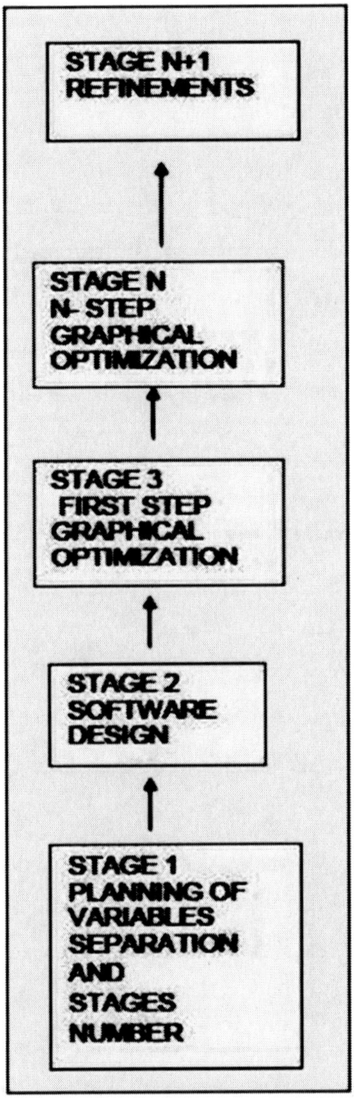

Sketch 1.- Brief of Interior Optimization Method stages. Any error at initial stages could cause significant bias in results. The algorithms are not general for a single program.

Usually, the software has to be adapted on the particular Nonlinear/Linear System(s) of the function(s).

To check complementarily the numerical results, both Fortran and F# subroutines in optimization were used. While Fortran proved to match the Graphical Optimization results, F# showed restrictions and obviously, limits that discarded this type of language for accurate results in nonlinear optimization. None of these numerical data were included in publication.

5. Pb COMPUTATIONAL GRAPHICAL-INTERIOR OPTIMIZATION RESULTS

The results for Pb are shown in Tables 1-2 and Figs. 1-2. The atomic mass of Pb is similar to Hg (Chapter I). However, alpha value is lower. Residuals can be considered acceptable, although not of high-precision accuracy, with short time consuming.

Results confirm useful utility of Equations 1-5, and show in graphics a sharp minimum line-zone at graphical optimization surface. Figs. 3-4 present GNU-OCTAVE Image Processing for comparison with Matlab in Figs 1-2.

NUMERICAL INTERIOR OPTIMIZATION RESULTS II		
Pb ISOTOPE TYPE RANGE (BY AMU)	K OPTIMAL	OBJECTIVE FUNCTION RESIDUAL (Chebyshev Optimization Norm)
208 (NATURAL)	31.62	0.5303
[204,208]	31.62	0.5303

Table 1.-The first stage of Pb Interior Optimization. In X and Y axes, are set the constant K and the parameter A that comprises Mα. That is, A= M^{α} in Fig 3. The values are taken almost from all references related to Pb critical temperature for superconductors. T_C value is fixed for natural Pb in simulations and at Z axis it is deployed the Objective Function taking absolute value. T_C for Pb is fixed because it is searched optimal values for all isotopes conditioned to the Tc of the natural Pb.

NUMERICAL INTERIOR OPTIMIZATION RESULTS III		
Pb ISOTOPE TYPE RANGE (BY AMU)	OPTIMAL ALPHA	FIXED Tc (Kelvin)
208 (NATURAL)	0.0404	7.197
[204,208]	0.0404	7.197

Table 2.- The second stage of Pb Interior Optimization. In X and Y axes Fig 4, are set the constant α and the parameter M. That is, previously was optimized $A= M^{\alpha}$.

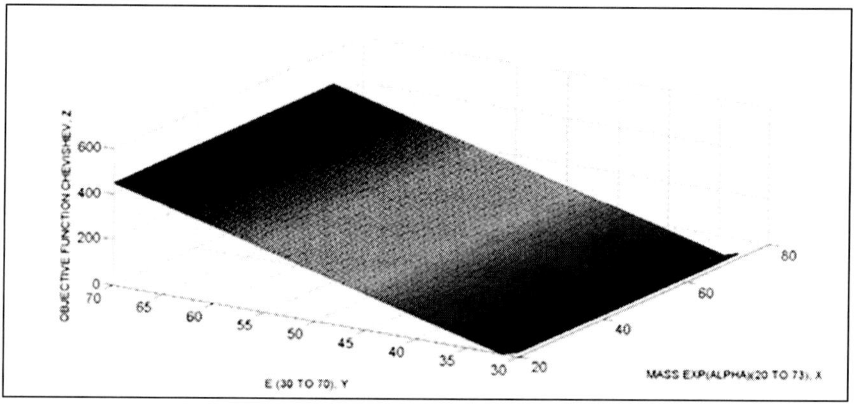

Fig. 1.-First stage of Pb Interior-Graphical optimization. It was determined the optimal K value and the optimal $A= M^{\alpha}$
value. Information and tools that can be used/obtained with imaging surfactal subroutines are detailed in [ch1-4].

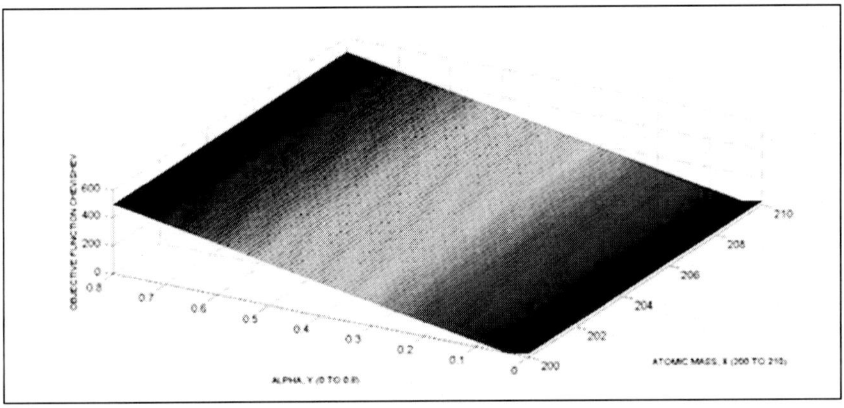

Fig. 2.- Second stage of Pb Interior-Graphical Optimization. It was determined the optimal M value and the optimal α value.

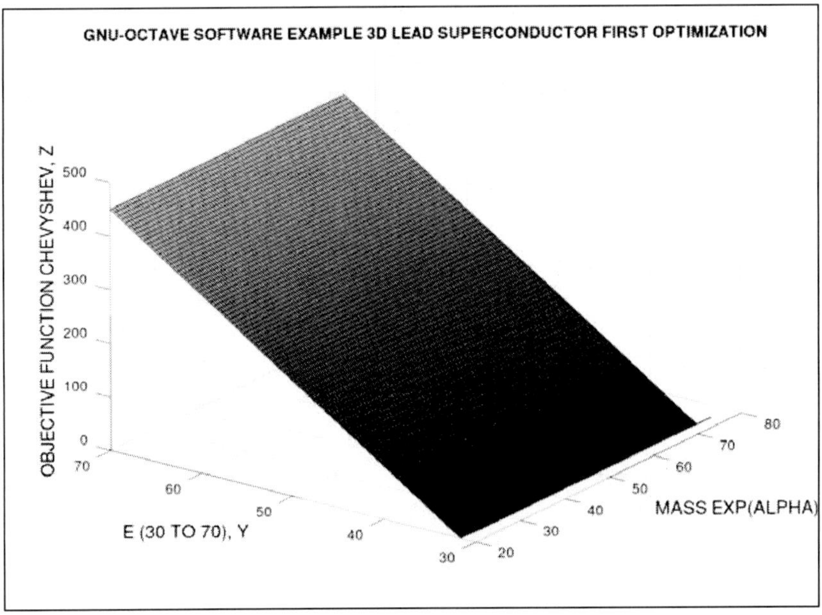

Fig. 3.-GNU-OCTAVE software example. First stage of Pb Interior-Graphical optimization. Image processing quality is acceptable and running time is slightly longer than Matlab. It was determined the optimal K value and the optimal A= M^{α} value. Information and tools that can be used/obtained with imaging surfactal subroutines are detailed in [ch1-4]. Note the acceptable imaging-processing quality.

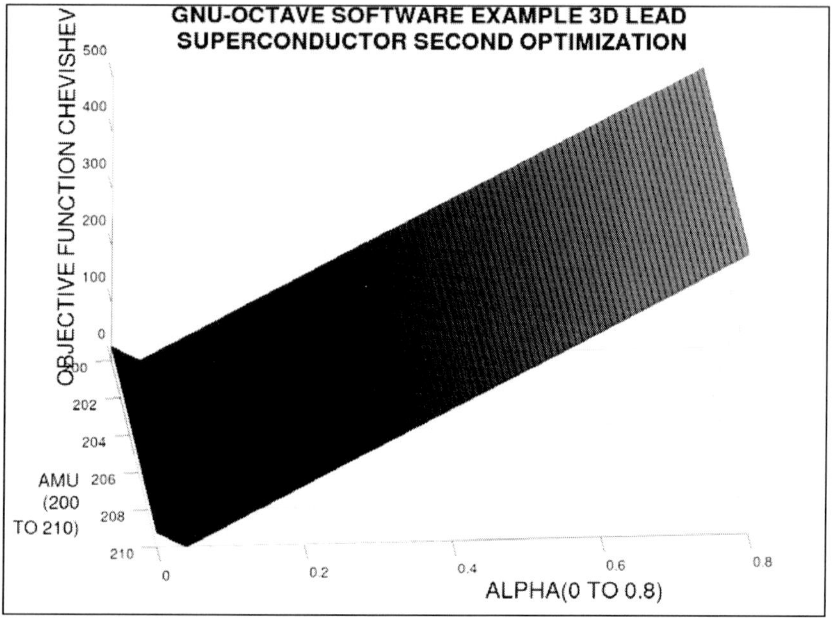

Fig. 4.-GNU-OCTAVE software example. Second stage of Pb Interior-Graphical optimization. Image processing quality is acceptable and running time is slightly longer than Matlab. The proof of Nonlinear System of equations unique solution can be seen at surface concavity. It was determined the optimal M value and the optimal alpha for A= M^{α} formula. Information and tools that can be used/obtained with imaging surfactal subroutines are detailed in [2-4].

6. Cr COMPUTATIONAL GRAPHICAL-INTERIOR OPTIMIZATION RESULTS

The results for Cr are shown in Tables 1-4 and Figs. 5-6. Atomic mass of Cr is not similar to Hg or Pb. Alpha value obtained is about 1/3 [Tables 1-4 for comparison]. Residuals are also good.

Residuals can be considered the best obtained in this study, with short time consuming. Results confirm useful utility of equations, and show in graphics a sharp minimum line-zone at graphical optimization surface [Appendix I].

NUMERICAL INTERIOR OPTIMIZATION RESULTS V		
Cr ISOTOPE TYPE RANGE (BY ATOMIC MASS)	K OPTIMAL	OBJECTIVE FUNCTION RESIDUAL (Chebyshev Optimization Norm)
52 (NATURAL)	33.33	0.1111
[50,52]	33.33	0.1111

Table 3.-The first stage of Cr Interior Optimization. In X and Y axes, are set the constant K and the parameter A that comprises M^{α}. That is, A= M^{α} in Fig 5. The values are taken almost from all references related to Cr critical temperature for superconductors. T_C value is fixed for natural Cr in simulations and at Z axis it is deployed the Objective Function taking absolute value. T_C for Cr is fixed in 3 K.

NUMERICAL INTERIOR OPTIMIZATION RESULTS VI		
Cr ISOTOPE TYPE RANGE (BY ATOMIC MASS)	OPTIMAL ALPHA	FIXED T_C (Kelvin)
52 (NATURAL)	0.3636	3
[50,52]	0.3636	3

Table 4.- The second stage of Cr Interior Optimization. In X and Y axes Fig 6, are set the constant α and the parameter M. That is, previously was optimized $A = M^{\alpha}$.

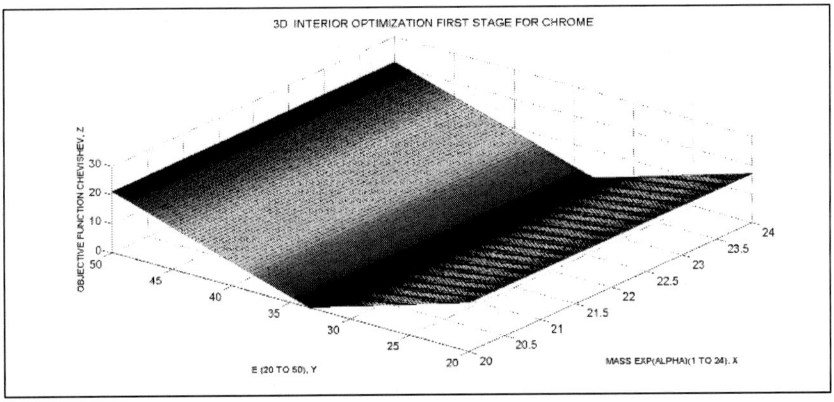

Fig. 5. - First stage of Cr Interior-Graphical optimization. It was determined the optimal K value and the optimal $A= M^{\alpha}$ value. Information and tools that can be used/obtained with imaging-surfactal subroutines are detailed in [ch1-4].

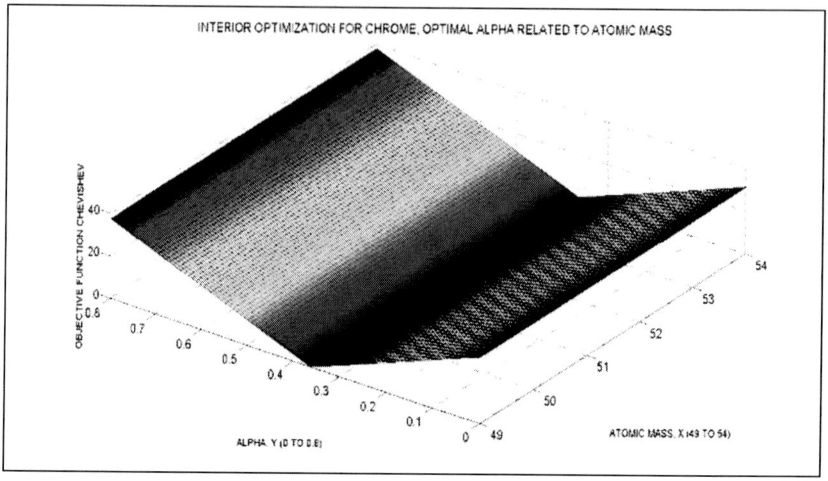

Fig. 6. - Second stage of Cr Interior-Graphical Optimization. It was determined the optimal M value and the optimal α value.

7. DISCUSSION AND CONCLUSIONS

The Interior Optimization Method base with algorithms and mathematical definitions that was explained/presented formally in [ch1], was developed and proven for several elements of Type I superconductors. The nonlinear systems of equations linked to the theoretical algorithmic-construction assertion that a System of Nonlinear Equations has not, in general, a unique solution was proven. The research was focused on superconductivity for Hg, Pb, and Cr, with acceptable results. GNU-Octave imaging processing software results be acceptable.

As a demonstration of the mathematical and practical engineering applications, electronics in superconductors equations were presented. Computational electronics simulations with 3D graphical optimization images were made correctly. Numerical results were explained in tables and schemes.

Numerical results match the usual values published in literature for Hg superconductivity effect. In alpha Hg value determination, the magnitude differs from the common 0.5 value set classically. Not always, however, 0.5 alpha value is calculated in all publications of the literature studied. The results for Pb show a small value of alpha, with good residual. The results for Cr show an excellent residual and an alpha value of about 0.4.

In Cr and Pb graphical Optimization, the proposed assertion of the author to prove that the uniquieness of Nonlinear Systems solution generally is comfirmed [Appendix I]

In summary, original method of Interior Optimization was developed with examples of practical electronics applications in superconductor electronics field. The non-existence, in general, of uniqueness of solutions for a system of nonlinear equations from Graphical Optimization Method was demonstrated. Applications in other areas of Electronic Physics, Physics, Bioengineering, Medical Radiation-diagnosis Technology and Engineering emerge from this Interior Optimization Method.

In Materials Tribology and Biotribology the usage of this method is similar with the Graphical Optimization technique that was published in previous contributions [2-4]. For example, as it was presented for medical devices implants optimization [23,27].

APPENDIX I

GRAPHICAL-NUMERICAL DEMONSTRATION OF VALIDITY OF ALGORITHM AND METHOD

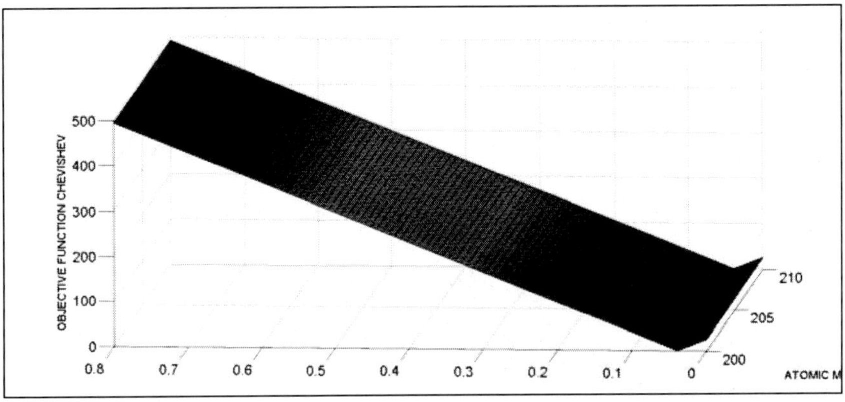

Figure 1. Appendix.-This image corresponds to stage 2 of Pb optimization. The validity of the simple BCS equation is sharply proven. It is shown a global minimum line, tangent of surface at z=0 level. In this tangent line, the alpha value remains constant for all different atomic masses of the isotopes. Mathematically this means that the algorithm works for several elements of Type I superconductors. That is,

$$M^{\alpha} T_C - E \approx 0$$

Usually a residual is found in optimization task. However, the initial BCS algorithms have been improved significantly with an extended mathematical background. This matter will be developed in upcoming contributions.

In the following Figure II. Appendix I, can be checked the same proof for Cr element. The non-uniquieness solution of a Nonlinear System of equations can be seen proven at the minimum tangent line observed at these graphics.

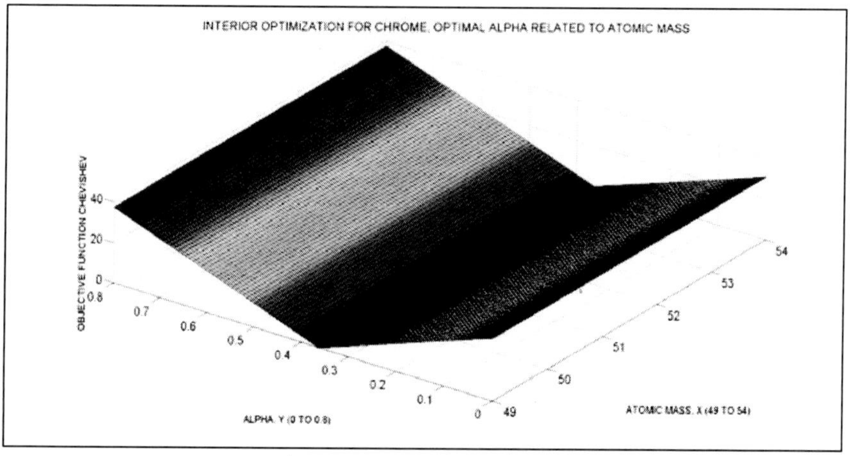

Figure 2. Appendix.-This image corresponds to stage 2 of Cr optimization. The validity of the simple BCS equation is sharply proven. The tangent of minimum values separates two parts of the interior-optimization surface. The reason is the programming design with Chevyshev L1 Norm taken in absolute value.

Original Article Reference

Casesnoves, F. "Advanced Interior Optimization Methods with Electronics Applications", International Journal of Scientific Research in Science, Engineering and Technology (IJSRSET), Online ISSN : 2394-4099, Print ISSN : 2395-1990, Volume 7 Issue 5, pp. 97-110, September-October 2020.
Available at DOI : https://doi.org/10.32628/IJSRSET207518
Journal URL : http://ijsrset.com/IJSRSET207518

CHAPTER III

INTERIOR DUAL OPTIMIZATION SOFTWARE ENGINEERING WITH APPLICATIONS IN BCS ELECTRONICS SUPERCONDUCTIVITY

ABSTRACT

Dual Optimization was the first approach towards multiobjective one in Interior Optimization [next Volume II to be released]. Interior Optimization software and algorithms-programming methods provide a computational tool with a number of applications. Theory and computational demonstrations/techniques were primarily shown in previous articles [3,4]. The mathematical framework of this new method, was also proven. The links among Interior Optimization, Graphical Optimization [Casesnoves, 2016-7], and classical methods in Nonlinear Equations Systems were developed. This paper is focused on software engineering with mathematical methods implementation in programming as a primary subject. The specific details for Interior Optimization computational-adaptation on every specific problem, such as engineering, physics, electronics physics, etc, are explained. Second subject is Electronics applications of software in the field of Superconductors. It comprises a series of new BCS Equation optimization for Type I superconductors, based on previous research for other different Type I superconductors previously published [3,4]. A new Dual Optimization for two superconductors is simulated. Results are acceptable with low errors and imaging demonstrations of the Interior Optimization utility.

Keywords : Interior Optimization Methods, Graphical Optimization, Systems of Nonlinear Equations, Electronics Superconductors.

1. INTRODUCTION

The design of software for optimization subject to a mathematical problem constitutes a rather difficult task. This inconvenient gets higher when there are many constraints and several objective functions/subfunctions. In previous publications, the Interior Optimization Method (IO) was presented and demonstrated [ch1-2]. Basically, it is a separation of variables method combined with Graphical Optimization 3D imaging/graphs.

Interior Optimization is not designed exclusively for finding out a specific global minimum, local minimum, partial-global minimum or variants of these singular parameters. Not always in Engineering and Physics is necessary to find optimal values for a group of equations conditioned to particular constraints of a laboratory/industrial/specific problem. Then, IO provides with the facility to select desired optimal parameters in subsequent stages ordered and set applying the method of separation of variables.

This computational contribution is focused on programming methods for IO and Dual IO. The Electronics applications in Superconductors constitute both a demonstrative part and a useful application at BCS superconductivity field.

In optimization with objective functions based on nonlinear systems of equations, or multiobjective-parameters constraints and other conditions, the task is rather difficult.

Specifically this occurs in Physics and Engineering nonlinear optimization problems [30]. Large-scale amount of empirical data for algorithms/models implementation have to be adapted/fitted to models that involve a high number of equations, nonlinear many of them, constraints added. One available useful method is Stochastic Optimization. These methods are based on random samples without taking all amount of data. They can be used to save time and computational effort when the amount of data is large.

In previous contributions, [2-4], Graphical Optimization was presented and applied for engineering mathematical modelling, materials erosion models [12], and physics of deformable solids dynamics. IO and Dual IO software implementation is difficult because the task involves a number of vectors and matrices to be implemented/ordered in the program code. This, specially in Dual IO was a hurdle sorted with Matrix Algebra Methods and is explained in Section III.

Graphical Optimization involves the objective function visualization in 2D and 3D.

This implies that multiparameter formulas initially cannot be easily set for the 3D Graphical Optimization, unless a choice of 2 or 3 variables is done among several ones. This apparent difficulty, was the base to develop Interior Optimization in previous contributions [2-4]. It is defined as Interior-Graphical Optimization/Dual Optimization Method [Casesnoves, 2018].

The method was intended to be proven and set, at first, on simple equations at Electronics Physics field, namely BCS algorithms [3,4]. Therefore, direct applications in electronics field for superconductors isotope-effect and critical temperature are presented.

In summary, this newly-improved-study shows an advanced contribution for the Interior Nonlinear Optimization Method previously presented in [2-4].

However, it is focused on programming software and computational 3D imaging results. Computational base is explained/proven at the Superconductors Type I area, Pt and Sn. The scheme of the method is shown, with a sharp proof in applications at this electronics field. The most important innovation is Interior Dual Optimization.

2. MATHEMATICAL METHOD

The problem approached was previously explained [ch1-2]. For software implementation IO and Dual IO is a system of nonlinear equations of several variables that has not, in general, unique solution. If the objective optimization function becomes a number of nonlinear functions, such as

$$f_i(\vec{x}) = f_i(x_{1i}, x_{2i}, \dots x_{ni})$$

(1)

for $i=1,\ldots.m$ and $n \neq m$ generally

subject to

$$a_{1i} \leq x_{1i} \leq b_{1i}$$
$$a_{2i} \leq x_{2i} \leq b_{2i}$$
$$a_{ni} \leq x_{ni} \leq b_{n}$$

(2)

or,

$$\begin{bmatrix} a_{1i} \\ \cdot \\ \cdot \\ a_{ni} \end{bmatrix} \leq \begin{bmatrix} x_{1i} \\ \cdot \\ \cdot \\ x_{ni} \end{bmatrix} \leq \begin{bmatrix} b_{1i} \\ \cdot \\ \cdot \\ b_{ni} \end{bmatrix};$$

(3)

These constraints are selected as simple here, in general could be more complicated. Then, setting equations members such as

$$f_i(\vec{x}) = f_i(x_{1i}, x_{2i}, \ldots x_{ni}) = 0$$

And Objective Function with L_1 Chebyshev Norm, for every f_i

$$\sum_{i=1}^{i=n} \left| f_i\left(\vec{x}\right) \right|_1 = 0 ;$$

[Casesnoves, 2018]

(4)

For programming in the following section, a nonlinear system of 100 equations was chosen, that is n=100.

3. IMPLEMENTATION OF METHOD IN SOFTWARE

The algebraic software developed for this contribution had various programming difficulties related to previous papers. To adapt the programs on Chebyshev L_1 norm for objective function with matrices in Dual Interior Optimization was complicated. Additional difficultiess were the algebra changes in the subroutines to get functional the program for surface Graphical Optimization visualization.

As it is detailed, [Sketch1], this requires mathematical changes-background. The software was developed in Matlab and is almost equivalent for FREEMAT also. Programming was selected to use subroutines that let rotate 3D images to obtain best visualization, [Sketch 2]. With this tool it was possible to see in images the sharp minimum zone to prove the mathematical statements. Freemat also offers this possibility in 3D surfactal implementation. There are more types of software that can perform this task properly.

Range for surfactal simulations Delanuay tiles was 100. In previous contributions, higher values were taken for high-accuracy [2-4,30]. However, optimization programs were designed for double precision, since were based on engineering software of extent number of previous publications [19-27].

INTERIOR OPTIMIZATION/DUAL OPTIMIZATION MATRIX ALGEBRA PROGRAMMING			
TYPE OF MATRIX	DIMENSIONS	SETTING	NUMERICAL SIGNIFICANCE
vector	[1,n]	to implement intervals	the length of the equations nonlinear system
vector	[1,n]	to implement intervals	to adapt the vector at matrix algebra (transpose)
square matrix	[n,n]	at programming body	for surfactal construction of OF
Non-symmetrical matrix	[n,m]	at programming body	intermediate matrix algebra calculations
Non-symmetrical matrix	[m,n]	at programming body	intermediate matrix algebra calculations

Sketch 1.-Brief of Interior Optimization Method Matrix Algebra programming. The particular characteristics for every type of matrix is explained. The setting on the program development is explained.

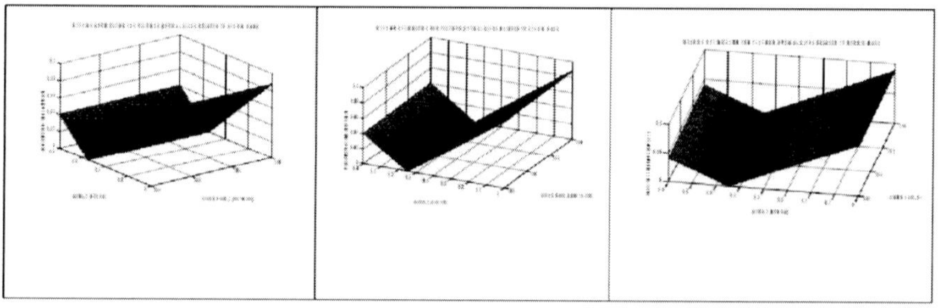

Sketch 2.-Brief sequence of global minima search with Graphical Optimization. The first image is given directly when running the software. Subsequent 3D images correspond to position variation of surfactal objective function to search global minima. Legends are not significant.

To check complementarily the numerical results, both Fortran and F# subroutines in optimization were used. While Fortran proved to match the Graphical Optimization results, F# showed restrictions and, obviously, limits that discarded this type of language for accurate results in nonlinear optimization. These numerical data, although significant for verification, were not included in publication.

4. APPLICATIONS IN ELECTRONICS

In this section a brief summary of Superconductivity towards the Transition Equation and the Isotope Effect is presented. The Superconductivity Theory framework is rather extent and with large mathematical background. It involves Quantum Theory, Molecular Chemistry, Materials Physics and other spezialities.

For the magnetic interaction of superconductor materials the Electromagnetic Theory is wide and essential—rather difficult and not the objective of this contribution . All these areas converge together in Superconductors theory and applications.

Superconductors developments in future stages of research involve a wide of theoretical and applied advances. Namely, in new materials compounds for Type II superconductors. Secondly, technical aspects for optimal implementation/usage could be developed. And the new technical/industrial applications for the future constitute an advance area to be worked out.

However, the most important industrial and scientific significances of superconductors are its essential property itself, resistivity cero at T_C . This fundamental property saves a loss of energy which constitutes the reason for their use/investigation.

What is useful to prove in Interior Optimization/Dual Optimization Method are the simple equations of Transition-Temperature-Critical and Isotope Effect.

The applications of Supeconductors in Nuclear Physics based devices , Medical Technology are described extensively in literature. Main of them are Hadrontherapy (Protontherapy and/or Carbontherapy) and Nuclear Magnetic Resonance (NMR). Protontherapy [24], is a high-precision dose delivery technique based on the advantage of Bragg Peak radiation distribution. Carbon also shows a satisfactory dose delivery distribution compared to photons. Hadrontherapy constitutes an improvement of dosimetry compared to classical IMRT. These applications are related to the area of magnetic fields interaction with superconductors. Following possible/probable applications in the future are in investigation and development.

The classical Critical Temperature Equation BCS for one isotope of a superconductors reads,

$M^{\alpha} T_C = K$;   (K=E in Figures)

(5)

where K (or E) is a constant, M Atomic Mass (AMU), and T_C is critical temperature (Kelvin). From the stage when the solid reaches critical temperature, the superconductivity effect begins. T_C varies for every isotope element within a defined mass element, although not excessively in magnitude.
This equation, although simple, remains useful. The evolution of this equation, for example, [2-4], involves extent mathematical background and mathematical-physics theory. Taking decimal of Neperian logarithms the equation (5) can be linearized for setting graphics related to different isotopes with corresponding atomic proper mass. The linearization objective is to guess fast information about isotopic superconductivity characteristics. In the literature [19-27] , it is frequent data analysis of BCS equation with linearization plots for comparisons.

The algorithms (1-5) were implemented with 100 equations and reads,

for (i=1, ...n) , i=100

$[M_i]^\alpha\ T_{Ci}\ -\ K_i\ = 0$, Tc fixed for simulations, and based on (4), it is set

$$\text{minimize,}$$
$$\left| [M_i]^\alpha\, T_{Ci} - K_i \right|$$
$$\text{for } i = 1,\ldots..n\ ;$$

$A_i = [M_i]^\alpha$ for first stage of optimization, e. g., Fig. 1

(6)

where i is the corresponding number within the range of simulations. For example, range of simulations for M in Hg [1-4] is,

$$M[\,Hg\,] \in [198.0\,,\,203.4]$$

(7)

In the same way, range of constraints can be seen in axes ranges of Figs 1-7. That is, 100 divisions in M range, (7) that comprise the variation of atomic mass for all metal isotopes within. Therefore, with equation (6), a nonlinear system of equations is set.

The variables are to get optimized. Extrapolation to large and complicated Electronic Physics equations, Electronics, Nuclear Physics applications, or any type of system can be guessed from these mathematical formulas.

5. Pt AND Sn COMPUTATIONAL GRAPHICAL-INTERIOR OPTIMIZATION RESULTS

This example/application deals with different superconductors compared to previous articles [3,4]. The results are shown in Tables and Graphics. The numerical applications are presented for Pt, and Sn, because of their similarity in atomic mass. However, Pt and Sn differ significantly in T_C and Isotopes distribution. Most of Sn isotopes correspond to AM 181, while Pt isotopes distribution shows almost the same percent for AMU 194 and 195.

Residuals can be considered acceptable, although not of high-precision accuracy. The time consuming for designing program is extremely short. The time for selecting optimal values is also very short with the Matlab graphical cursor. In engineering practice, this is a significant advantage that gives many options for fast on-site superconductivity data management at laboratory. Results confirm useful utility of equation (5).

Pt NUMERICAL INTERIOR OPTIMIZATION RESULTS I		
ISOTOPE TYPE INTERVAL OF ATOMIC MASS (AMU) AND T_C	PARAMETER	INTERVAL APPLIED FOR PROGRAMMING
$M \in [190,198]$ $T_C = 0.0019$ K	$A = M^{\alpha}$	[1.0,68.7594]
	$M^{\alpha} T_C$	[0.0019,0.1306]
	E	[0.0019,0.1306]
	ALPHA	[0.001,0.8]

Table 1.-The first stage of Interior Optimization. In X and Y axes, are set the constant K and the parameter A that comprises M^{α}. That is, A= M^{α} in Fig 1, X axis. The values are taken almost from all references related to Pt critical temperature for superconductors. T_C value is fixed. At Z axis it is deployed the Objective Function taking absolute value. It tooks about 15 minutes. Programs construction were performed in 2 hours based on previous software with matrix algebra modifications.

Pt NUMERICAL INTERIOR OPTIMIZATION RESULTS II		
OPTIMIZATION PARAMETER	NUMERICAL RESULT	RESIDUAL
E (K)	0.0901	0.0869
ALPHA	0.5495	0.0002

Table 2.- The second stage of Interior Optimization. In X and Y axes, are set the constant α and the parameter M. That is, previously was optimized $A = M^{\alpha}$. The values are taken almost from all references related to Pt critical temperature for superconductors. The Objective Function is taken in absolute value. These values were taken with the cursor in the most easy and simple way. It took about 10 minutes.

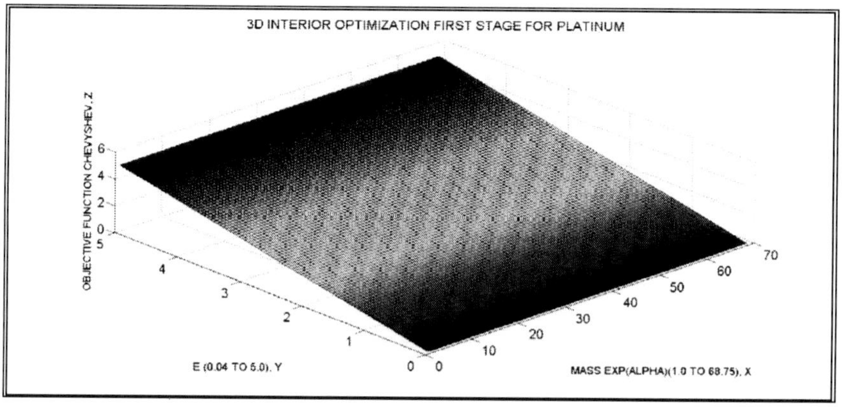

Fig. 1.-First stage of Pt Interior-Graphical optimization. It was determined the optimal K (E) value and the optimal $A = M^{\alpha}$ value. Information and tools that can be used/obtained with imaging-surfactal subroutines are detailed in [2-4,19-27]. The absolute value was taken at Chebyshev objective function.

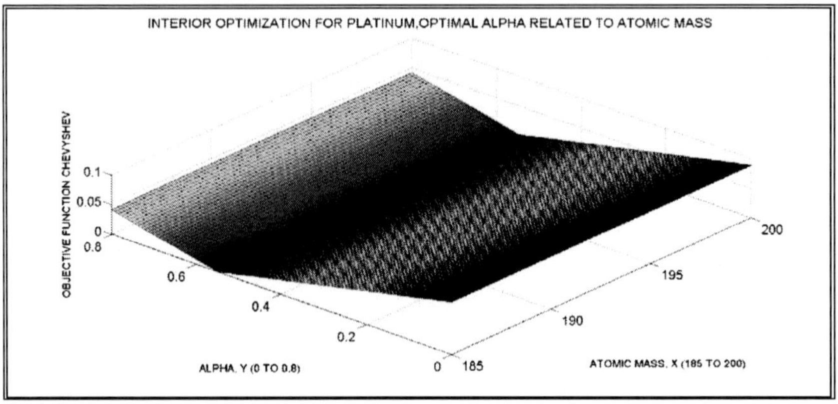

Fig 2.- Second stage of Interior-Graphical Optimization. It was determined the optimal M value and the optimal α value. Information and tools that can be used/obtained with imaging-surfactal subroutines are detailed in [2-4,19-27].

Sn NUMERICAL INTERIOR OPTIMIZATION RESULTS I		
ISOTOPE TYPE INTERVAL OF ATOMIC MASS (AMU) AND T_C	PARAMETER	INTERVAL APPLIED FOR PROGRAMMING
$M \in [175,185]$ $T_C = 3.72$ K	$A = M^{\alpha}$	[2.8094,65.1234]
	$M^{\alpha} T_C$	[10.4508,242.2592]
	E	[10.4508,242.2592]
	ALPHA	[0.2,0.8]

Table 3.-The first stage of Sn Interior Optimization. In X and Y axes, are set the constant K and the parameter A that comprises M^{α}. That is, A= M^{α} in Fig 1, X axis. The values are taken almost from all references related to Sn critical temperature for superconductors. T_C value is fixed. At Z axis it is deployed the Objective Function taking absolute value. All these values were taken with the cursor in the most easy and simple way. It took about 15 minutes. Programs construction were performed in 2 hours based on previous software [19-27] with matrix algebra modifications.

Sn NUMERICAL INTERIOR OPTIMIZATION RESULTS II		
OPTIMIZATION PARAMETER	NUMERICAL RESULT	RESIDUAL
E	36.11	0.09618
ALPHA	0.2667	0.09742

Table 4.- The second stage of Sn Interior Optimization. In X and Y axes, are set the constant α and the parameter M. That is, previously was optimized A= M^{α}.

The values are taken almost from all references related to Sn critical temperature for superconductors. The Objective Function is taken in absolute value. These values were taken with the cursor in the most easy and simple way. It took about 12 minutes.

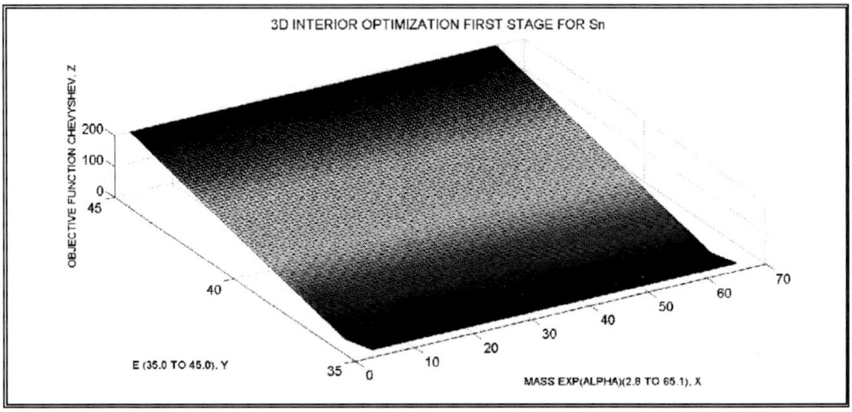

Fig. 3.-First stage of Sn Interior-Graphical optimization. Line of contact for global minima is not at center. It was determined the optimal K (E) value and the optimal A= M^{α} value. Sn is a more stable metal related to isotopes mass distribution. Information and tools that can be used/obtained with imaging-surfactal subroutines are detailed in [ch1-3] . The absolute value was taken at Chebyshev objective function.

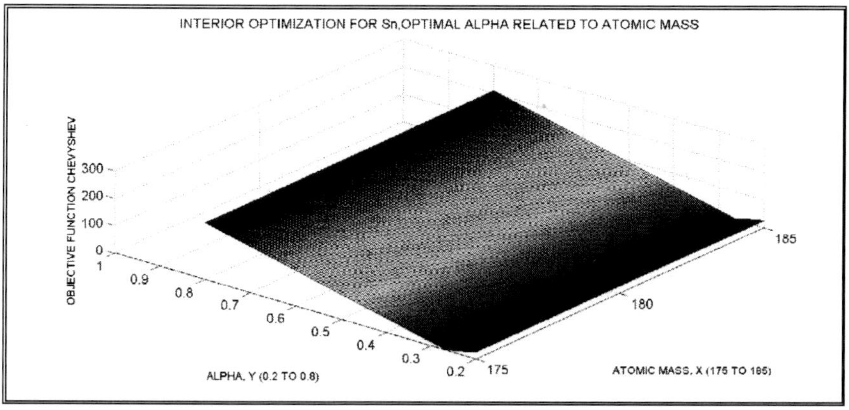

Fig 4.- Second stage of Sn Interior-Graphical Optimization. It was determined the optimal M value and the optimal α value. Information and tools that can be used/obtained with imaging-surfactal subroutines are detailed in [ch1-3].

6. DUAL Pt AND Sn COMPUTATIONAL GRAPHICAL-INTERIOR OPTIMIZATION SIMULATION RESULTS

This section deals with a theoretical simulation of BCS equation for two materials, Pt and Sn. They have similar atomic mass, although the isotopic distributions are different.

It is extensively known, that Pt has two isotopes with percentage similar, about 33% (194 and 195 UAM). On the contrary, Sn shows about 100% isotopic distribution at 180 atomic mass isotope (UAM). The objective is to demonstrate the utility of Interior Optimization for several theoretical and experimental situations.

The purpose consists in a dual optimization with IO for both materials in three stages—namely, separation of variables. It is supposed, for instance, that at laboratory cables with powder of these two materials are being manufactured.

That is, cables can be bent, since it is powder of both materials mixed. Therefore, cables could adapt on any convenient lab device. It is also supposed that in the mixture there is no interaction between the two different powered metals. And also there is absence of Electromagnetic fields at the set up device. Besides, other application for SCMTL is explained in Chapter IV.

The IO application in this case, involves three subsequent stages. These steps are compulsory since Pt and Sn have different T_C . It is searched as follows,

Optimal T_C for both metals within an optimal range of T_C

Optimal E (K in some graphs/formulas of previous papers) in BCS equation for both metals

Optimal alpha value for BCS equation applicable on both metals

Pt has a T_C of 0.0019 K, while Sn T_C is 3.72 K.. In simulation, is is taken a range $T_C \in$ [0.0001,0.0019]. Therefore, once reached this range of temperatures, it is assured that the superconductivity effect will happen in both elements. The software engineering programming is more complicated and was done in double precision. Matrix Algebra changes are frequent and rather complicated. It was mandatory a strict order in programming sentences to avoid problems when running the program [Casesnoves software, 2020], and save functional time.

In Tables 4-5, numerical results are presented. In graphs 5-7, the sharp IO in three stages is shown and clearly demonstrated. Table 6 is a brief of programming stages. Explanations of results are included at everyone. Just remark that this is a theoretical simulation based on rational experimental situations in lab.

That is, the necessity of bending cables of powered-superconductors for an eventual device settings/manufacturing.

Residuals can be considered acceptable at this first dual IO. The time consuming for designing program is about 5 seconds. The time for selecting optimal values is not short with the Matlab graphical cursor as there are 3 stages. In engineering practice, this simulation is a significant advantage that gives many options for fast on-site superconductivity dual-data management at laboratory. Results confirm useful utility of equation (5).

DUAL NUMERICAL INTERIOR OPTIMIZATION RESULTS V Pt AND Sn		
STAGE OF OPTIMIZATION	OPTIMAL RESULTS FOR PARAMETERS	OBJECTIVE FUNCTION RESIDUAL (Chebyshev Optimization Norm)
FIRST STAGE	K=0.1317 $M^{\alpha} T_C$ =0.1388	0.008302
SECOND STAGE	T_C = 0.0005909 M^{α} =69.31	2.48e-6
THIRD STAGE	Alpha=0.7 M ϵ [175.0,200.0] (UAM)	0.1076

Table 5.-Numerical results of Dual Interior Optimization for Pt and Sn together. Residuals can be considered acceptable. The value of Alpha is applicable for both metals at BCS equation. Optimal isotope mass that can fit better in BCS equation is 189.7 UAM for first stages. At third stage, all isotope masses, both Pt and Sn, M ϵ [175.0,200.0] are optimal with good residual, and alpha value 0.7.

DUAL NUMERICAL INTERIOR OPTIMIZATION RESULTS II		
STAGE	PARAMETER FIXED	OPTIMAL PARAMETER DETERMINED
FIRST STAGE	E (K) TENTA-TIVE	E (K)
SECOND STAGE	E (K) FIXED	Tc
THIRD STAGE	Tc E (K)	ALPHA

Table 6.- The development of optimization of parameters in Dual Interior Optimization. These stages correspond with Figs 5-7. The numerical results for these parameters are in Table 5.

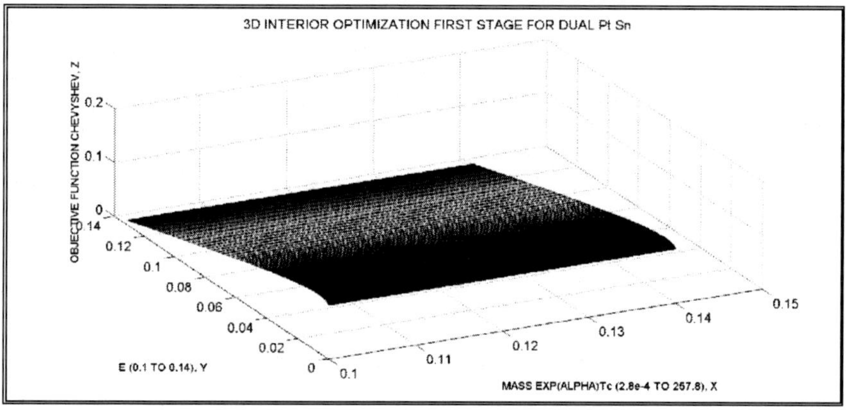

Fig. 5.-First stage of Interior-Graphical Dual Optimization. It was determined the optimal E value and the optimal A= M^{α} T_C values. The interval selected for T_C is [0.0001,0.0019]. Intervals are in legends of graphs 5, 6, and 7. Numerical results are in Table 5.

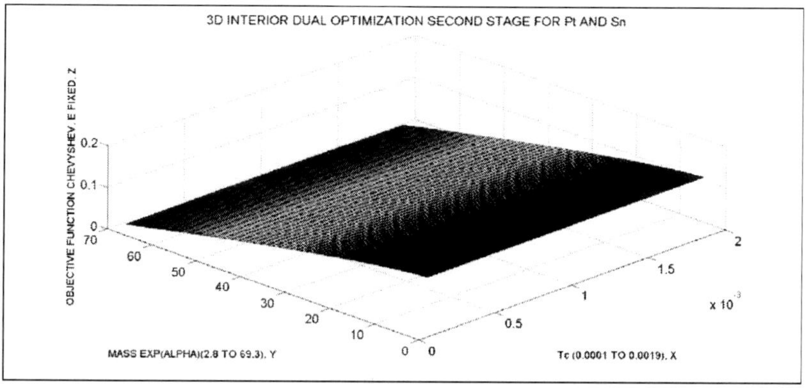

Fig 6.- Second stage of Interior-Graphical Dual Optimization. It was determined the optimal E value was fixed from first stage, E=0.1. Intervals are in legends of graphs 5, 6, and 7. Numerical results are in Table 5.

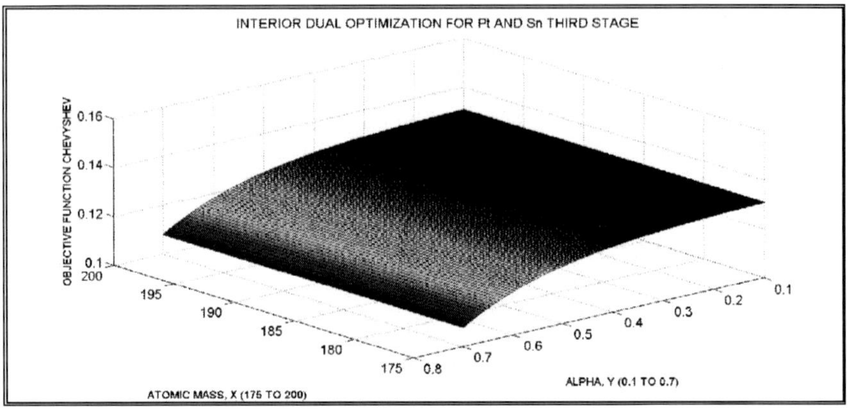

Fig 7.- Third stage of Interior-Graphical Dual Optimization. It was determined the optimal Alpha value. Intervals are in legends of graphs 5, 6, and 7. Numerical results are in Table 5. Note the smooth curvature of the surface. This mathematically implies that the OF is differentiable, derivable continuously. The homogeneity for all the isotope atomic masses is excellent with alpha value of 0.7.

7. DISCUSSION AND CONCLUSIONS

The aim of this paper was different from previous contributions. It is focused in software engineering techniques to implement the Interior Optimization. Therefore, the superconductors applications for Pt and Sn are confirmation examples of the method previously demonstrated.

However, both Pt and Sn results can be considered acceptable. Dual simulation, [Casesnoves, 2019-2020] is a new tentative perspective for theoretical applications in Electronics Physics. The basic stages of IO were detailed with simulations and Graphical Optimization 3D imaging. In Dual Interior Optimization, even some refinements were implemented.

Therefore, Interior Optimization Method base with algorithms and mathematical definitions that was explained/presented formally in [ch1-2], was developed and proven for several elements of these different Type I superconductors. One important feature of this method is its speed in running time even working with double precision.

The utility, although simple, of the BCS equation was proven for Pf and Sn. The Matrix Algebra programming is rather more difficult compared to previous contributions. This occurs in the Dual IO program design. The difficulties were sorted by using several combined code-sentences.

As a demonstration of the mathematical and practical engineering applications, electronics in superconductors equations were presented. Computational electronics simulations with 3D graphical optimization images were made correctly. Numerical results were explained in tables and schemes.

Numerical results for alpha in Pt are about ½. For Sn, the result of around 0.3 differs slightly from the literature value of 0.44. Dual alpha for both metals fits at 0.7 with a very good residual. In Dual Interior Optimization, the optimal value $T_C \approx 0.0006$ K guarantees the superconducting effect for Pt and Sn at the same time.

In summary, original Software Engineering method of Interior Optimization/Dual Optimization was explained with a number of examples. Software techniques are supported by numerical results and 3D images/graphs. Practical Electronics applications in Superconductor Electronics field BCS equation are presented. The simulation of Dual Interior Optimization was carried out as much realistic as possible. Nuclear Physics applications in Radiation Therapy (Hadrontherapy), Diagnostic Imaging (NMR) emerge from these findings. Applications in other areas of Electronic Physics, Physics, Bioengineering, Medical Radiation-diagnosis Technology, Tribology, and Engineering come from this Interior Optimization/Dual Optimization Method.

Original Article Reference

Casesnoves, F. Interior Dual Optimization Software Engineering with Applications in BCS Electronics Superconductivity.
Asian Journal of Mathematical Sciences (AJMS).
AJMS/Jul-Sep-2020/Vol 4/Issue 3 . 2020.

CHAPTER IV

SUPERCONDUCTOR MULTIFUNCTIONAL TRANSMISSION LINE FUNDAMENTALS

ABSTRACT

Primary Superconductors Multifunctional Transmission Line (SCMTL) engineering basic project built on High Temperature Superconductors (HTSCs) Interior Optimization (IO) and Molecular Effect Model (MEM) previous research series is explained and detailed. Results comprise fundamental design of SCMTL and functional applications with several HTSCs materials. Advantages and inconvenients of SCMTL project are evaluated. A number of applications are presented. Transmission Lines (TLs) and optimization in several electric power installations types are shown. In Appendix, recent Artificial Intelligence MEM results that can be applied in SCMTLs are presented and demonstrated.

Keywords : Interior Optimization (IO) Methods, Graphical Optimization, Isotope effect (IE), Critical temperature [Tc], Inverse Least Squares (ILS), Electronics Superconductors (SC), High-Temperature Superconductors (HTSC), [Tl-Sn-Pb-Ba-Si-Mn-Mg-Cu-O] Molecular HTSC Group, Molecular Mass (MO), Molecular Effect Model (MEM), Transmission Lines (TL), Power Transmission Lines (PTL), Superconducting Multifunctional Transmission Line (SCMTL).

1. INTRODUCTION

A superconductor can be defined as any material type whose electrical resistance is approximately null under specific thermodynamic and electromagnetic conditions. The essential thermodynamic conditions needed to reach the superconductivity state are given by a critical temperature T_C, beyond which, toward lower temperatures, a superconductivity effect takes place and interactions with magnetic fields constitute an important modifying factor. The T_C magnitude is around absolute zero Kelvin for conventional superconductors and approximately 100 degrees higher for high-temperature ones. Apart from this crucial condition, there are other physical oncs. Namely, the maximum critical current, lower

critical magnetic field H, and upper critical magnetic field H. Other factors are pressure and resistivity [ch4]. In general, the large variety of models and formulations within the theory. Following a series of superconductors and high-temperature superconductors series, this study presents the Superconducting Multifunctional Transmission Line (SCMTS) basic concepts and design. An SCTMS is defined [ch4] as follows,

Definition I.-A Superconducting Multifunctional Transmission Line (SCMTL) is a multi-operative transmission line adaptable on electrical/magnetic power energy supplies/demands optimization in function of environmental/artificial temperatures and electrical power parameters [Casesnoves, 2021] . PTL design is mainly based on three science and engineering foundations. Namely, Power Engineering, Electromagnetism Physics Maxwell Theory and Electromagnetic Theory, and Applied/Theoretical Mathematics. Statistical

Optimization takes also an important role of the advances in this field. In general, the design, manufacturing and maintenance of a PTL is complex and depends on multiple factors.

For the objectives of this study, the most important are, materials conductors, overheating and overloading, insulation and coatings, electrical and magnetic parameters, and environmental ones. Additionally, for conventional PTL, corrosion, biological contamination, lubrications, plating and aging/life-expectancy are also significant.

The PTLs common materials are aluminum, copper and steel. Their electrical resistivity difference, compared to any superconductor, either Type I or II, is enormous in magnitude.

Currently, energy demands are increasing in industrial world significantly, and the prospective demands expectations for next times show be very high [ch4]. This implies that innovative use of superconductors could reduce energy losses and decrease the electrical power production and consumption costs.

The basic property of superconductors is not the main advantage of a SCMTL. A significant problem in PTLs is the overloading and overheating. Overloading could be sorted by far with superconductor materials. Overheating also, because of the functional-operative temperatures of superconductors. Cooling cost, however, constitutes an economical strain. Moreover, optimization of power demands could be dealt better with combination of superconductors in a SCMTL design.

A SCMTL is a cylinder-shaped structure whose internal structure has several encircled layers. The layers and central line could be formed by conducting, superconducting, or HTSCs materials. It could be coaxial or pair line also. The innermost layer may be formed by a conventional conductor—for electrical power engineering standard transmission. From inner structure to outwards one, several stratums of divided/different HTSCs and/or superconducting circular layers/covers could be installed.

Cooling covers/stratums are between any subsequent superconductors or HTSCs classes. From inner direction to outwards one, Figure 1, every HTSC magnitude of T_C is decreasing/increasing according to design conditions. For the most external superconductor layer has the maximum/minimum T_C magnitude. Consistent with electrical power engineering requests, the operational usage could be from one to all superconducting layers working at the same time or in successive intervals. Multiple optimization options are possible. Even the conventional inner conductor could be working singularly according to electronics-thermodynamical and/or environmental energy constraints at any time.

Thus, the innovation of this research is to present a new primary design for SCMTLs. Fundamentals are explained and technically justified. Applications, advantages and inconvenients are evaluated in the project from an engineering and electronics physics criteria. In summary, SCMTL offers theoretically prospective applications both for electrical power engineering and wave TLs.

2. ELECTRONICS PHYSICS INVENTION BASIS

This SCMTL line has capability to transmit electrical power at a wide range of temperatures, depending on the engineering resources and power energy demand . The energy demands at industrial countries and increased electrical power use has increasing prospective use for future [ch4]. Both for electrical power transmission and electromagnetic waves TLs energy losses represent a severe difficulty for operational/economical optimization. Minimization of the electrical resistance is an optimal possible method to avoid electrical power loss transmission lines diminution. HTSCs and standard superconductors materials could work out that hurdle. If the engineering design/operative-functionality perform properly. In addition, the HTSCs and superconducting materials manufacturing/maintenance costs should be moderate.

In power SCMTL, the electrical power savings are principally linked to significant materials resistance diminution. Figure 1 shows the basic structure of a SCMTL. The main parts are:

Cylinder-shaped structure encircled layers of conducting and HTSCs materials.

Conventional structure, coaxial, parallel or pair line (not applicable).

Optional superconducting materials resistance distribution: from lower to higher or inversely T_C magnitude, from inner part towards external part.

Cooling layers optimal distribution.

Optional operative functions. The use of a number of HTSCs and/or superconductors combined with/without conventional transmission material(s).

Thermodynamical multiobjective optimization for all T_C magnitudes.

Magnetic fields created by the superconductors and/or HTSCs layers while power conducting could be synergic also and increase the efficiency by modifying better the T_C magnitude.

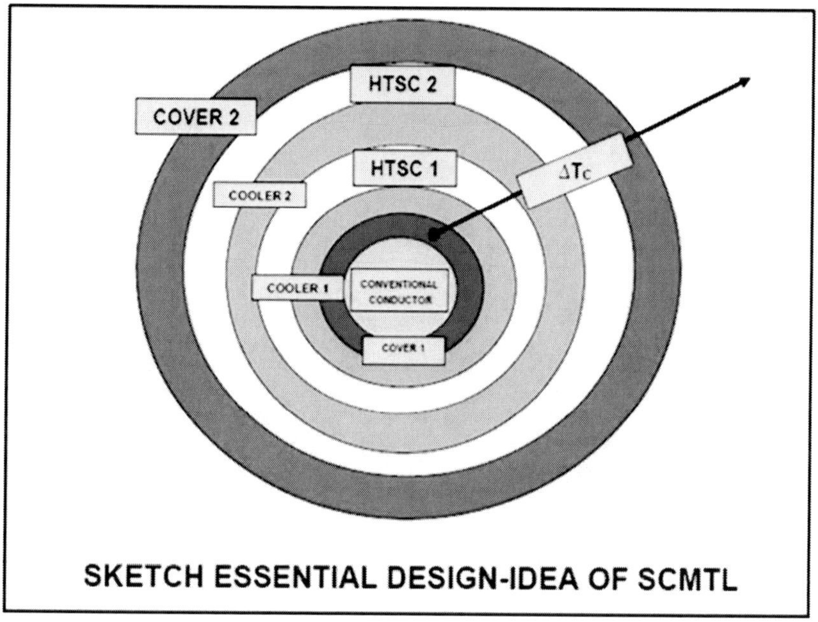

Figure 1.- From [ch4], basic design of a SCMTL. HTSC1 , first superconductor type, HTSC2 the second one with higher T_C . Direction of T_C increase [arrow] marked with arrow. Conventional conductor at center. This is a simple idea-design.

1.1. Interior Optimization of Isotope Effect and Molecular Effect for optimal design applications

In former publication series [ch1-3], Interior and Graphical Optimization Methods were developed, together with Molecular Effect Model/Hypotheses (MEM) for Types I and II superconductors. Further, MEM constitutes an important prospective technique for HTSCs combinations in object. These extensive dataset can be applied for optimal T_C design at SCMTLs. Figures 2-3 show examples of 2D and 3D MEM and Interior and Graphical Optimization of superconductors and HTSCs. Figure 4 proves the utility of Genetic Algorithms Software Engineering for HTSCs. Both 2D/3D/4D Interior/Graphical Optimization and Isotope/Molecular Effects models could be suitable for the presented project.

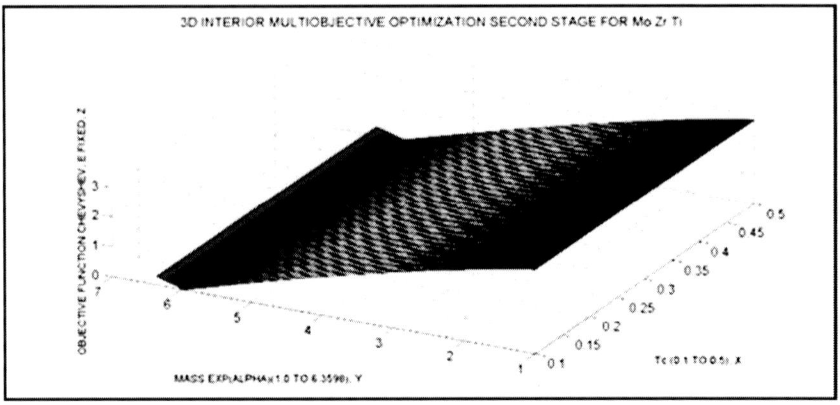

Figure 2 .- From [ch4 refs], second stage of Interior-Graphical Multiobjective Optimization for Mo Zr Ti. It was determined the optimal E value was fixed from first stage, E=2.909. It was selected the E1 value of Table 3. Intervals are in legends of graphs 4, 5, and 6.

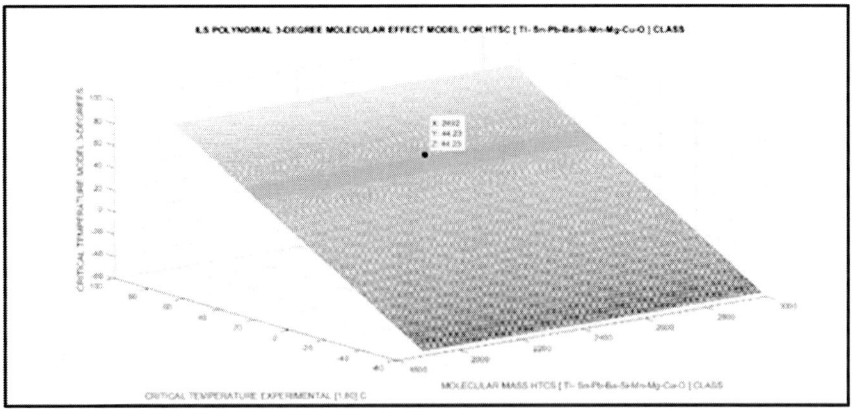

Figure 3.- From [ch4 refs], 3-Degree polynomial MEM 3D graph showing model T_C prediction and 3D experimental data which are approximately equal. The HTSCs class for this software graphical optimization is [Tl-Sn-Pb-Ba-Si-Mn-Mg-Cu-O] . Molecular HTSC Group Numerical example coincidence between experimental and MEM data is marked.

Note the dark zone at surface that means the transition between negative and positive T_C for the MEM. This matches the 2D results of Figures 3-5. At Z axis MEM data with imaging processing method 2. Matrices for 3D Graphical Optimization are [300 x 300].

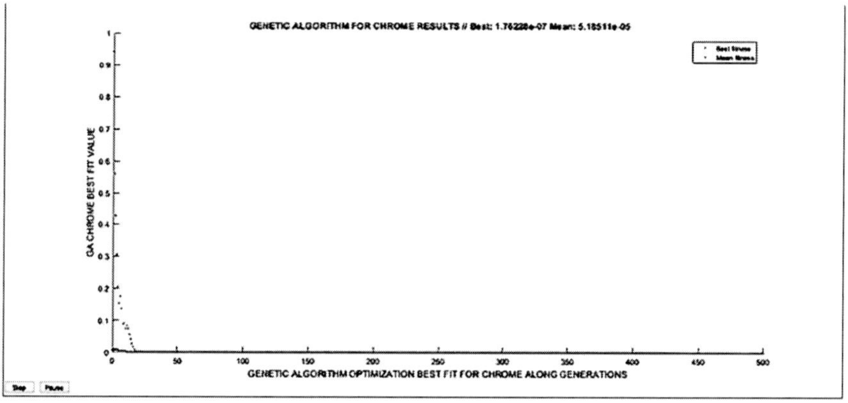

Figure 4 .- From [ch4 refs], example Interior Optimization of Chrome Isotopes with Genetic Algorithms. The Chrome GA results (first stage optimization) for best fit along has generations (500). Results show be accurate [ch4 refs, next Volume II].

2. POWER ENGINEERING APPLICATIONS

The prospective applications of SCMTL are theoretically wide. There are TLs for waves, generally, and electrical power TLs. All of them get advantages theoretically from SCMTLs. However, the most useful to save energy is the electrical power TLs. In waves TLs, distance from surface of HTSC, is inversely proportional to power loss. This difficulty is sharply attenuated by the superconducting effect. Table 1 shows a functional analysis of SCMTLs and Table 2 presents their prospective applications.

ANALYSIS OF FUNCTIONS OF SCMTL	
ADVANTAGES	**INCONVENIENTS**
High voltage/intensity performance	Not very long connections for high-voltage and current demand at Supercomputers Installations. Cooling systems design/maintenance compulsory/necessary
Optimization for energy demands	High reparation/maintenance costs
Very high energy cost reduction	Magnetic/electrical field environmental contamination and risks. Costs for shielding.
Additional use of magnetic fields created by current	Complicated optimization to synchronize all the different superconductor layers functions
Sorting Overheating and Overloading problems	A number of design/manufacturing difficulties proportional to length of SCMTLs

Table 1.-Primary evaluation of SCMTL project advantages and inconvenients on applications for power engineering and electronics physics.

BRIEF OF SCTML APPLICATIONS		
USAGE FIELD	**ENGINEERING**	**COMMENTS ADVANTAGES/INCONVENIENTS**
High performance computational laboratories/ Supercomputers	Not very long connections for high-voltage and current demand at Supercomputers installations	Possible with technical difficulties. Very useful for varying high-electrical power demand of Supercomputing installations.
Medical Technology Apparatus	MRI or similar medical devices that create high magnetic fields due to very high electrical intensity	Possible with technical difficulties. The same reasons than above.
Transport based on Magnetic Fields levitation tracks. [levitation train, manly]	Magnetic field levitation based on superconductor that create very high magnetic fields	Much more difficult
Aerospatial Stations	High electrical intensity and/or magnetic fields required for Spatial Station technology	Much more difficult

Table 2.-Primary fundamental applications of SCMTLs. Just remark that those are primary applications .

3. DISCUSSION AND CONCLUSIONS

The objective of the contribution was to present the basic engineering project for a SCMTL. Foundation ideas and prospective applications are shown. The base of superconducting effect through increasing magnitudes of T_C and layers distribution was presented.

Previous research series of Interior and Graphical Optimization for Isotope Effect and Molecular Effect in SCs and HTSCs were explained for project design [ch4].

The project innovation is given by energy demands losses, budgets reduction, and power demands optimization. Multiple applications, Table 2, constitute a number of plusses.
Advantages for electrical power engineering are the savings of both resistance (impedance with attenuation factors) and other losses of conventional PTLs related to electrical resistance. Inconvenients are given by the manufacturing, repairing, and maintenance costs.

The Artificial Intelligence applications for SCMTLs design correspond to next Volume and articles.

In brief, an initial SMTL project was explained at basic stage. Applications for electrical power engineering are wide with several prospective substantial energy production increase. Diverse uses for several technology fields emerge from the project.

4. APPENDIX MOLECULAR EFFECT MODEL ADVANCES

In this section a few examples of MEM to be published are shown for Artificial Intelligence Evolutionary Algorithms MEM optimization for high temperature superconductors cuprates HTSCs [Hg-Ba-Ca-Cu-O] .

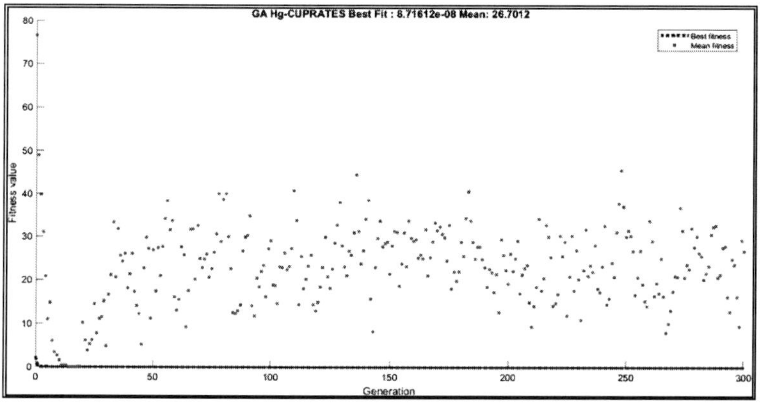

Figure 1.-Best Fit, with 5 optimization parameters, the most important chart of GA method. Generations Number is 300. As it was found, the MEM shows an approximate 2D parabolic shape in 3D charts.

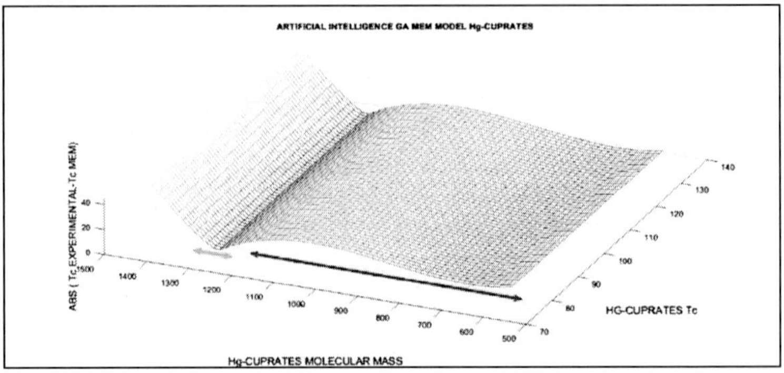

Figure 2.- Artificial Intelligence Optimal MEM sector for Hg-Cuprates HTSCs class. Generations Number is 300. Green arrow zone, the almost exact fit, red arrow, inset, shows the approximately acceptable fit. Hg-Cuprates MEM shows be the best at this stage. Model fits approximately well between molecular weights interval [500 , 1400] for all values of T_C.

Original Article Reference

Casesnoves, F. Superconducting Multifunctional Transmission Line Fundamentals
International Journal on Recent Technologies in Mechanical and Electrical Engineering
ISSN: 2349-7947 Volume: 9 Issue: 3 . 2022
IJRMEE | September 2022, Available @ http://www.ijrmee.org

1. NOTATION AND ACRONYMS

Notation is referred to mathematical symbols. Acronyms is related to abbreviations.

1.1. Notation

Symbol	Meaning
R	Real numbers space (Complex numbers space is not used in book)
$=$	Equal to
$\leq$	Less or equal to
$\geq$	High or equal to
$<$	Strictly less than
$>$	Strictly higher than
$\pm$	More/minus (usually used for errors)
$(\bullet , \bullet)$	Open Interval
$[\bullet , \bullet]$	Closed Interval
$\in$	Belonging to (usually an interval)
$\sum$	Summatory
$\lVert \bullet \rVert$	L_2 Norm
$\lvert \bullet \rvert$	L_1 Norm (for multiobjective Optimization denoted as Cheby-shev Norm)

$\vec{X}$	Vector ϵ R
X	Vector Norm ϵ R
$f(\vec{X})$	Vectorial function (several functions and/or variables)
M	Atomic Mass (usually isotopes) or Molecular Mass

1.2. Acronyms

Acronym	Meaning
SC	Superconductor in general
HTSC	High Temperature Superconductor
T_C	Critical Temperature (Kelvin [K] or Celsius [C])
ILS	Inverse Least Squares Optimization Method
SCMTL	Superconducting Multifunctional Transmission Line
IE	Isotope Effect
AMU	Atomic Mass Unit
MO	Molecular Mass
MEM	Molecular Effect Model
TL	Transmission Lines
PTL	Power Transmission Lines
OF	Objective Function
IO	Interior Optimization
ch1	In references, it means references of article of Chapter I
ch2	In references, it means references of article of Chapter II
ch3	In references, it means references of article of Chapter III
ch4	In references, it means references of article of Chapter IV

phd	In references, it means references of Author's Doctoral Thesis [12]

2. GENERAL REFERENCES LITERATURE

1. Vora, A.M. (2008). Modified Transition Temperature Equation for Superconductors. Chin. Phys. Lett. 2008, 25, 2162. [CrossRef]

2. Casesnoves, F. (2020). Interior Optimization Methods with Electronics Applications. Int. J. Sci. Res. Sci. Eng. Technol. 2020, 7, 428–436. [CrossRef]

3. Casesnoves, F. (2020). Advanced Interior Optimization Methods with Electronics Applications. Int. J. Sci. Res. Sci. Eng. Technol. 2020, 7, 97–110. [CrossRef]

4. Casesnoves, F. (2020). Multiobjective Interior Optimization Computational Methods for Electronics BCS Superconductivity. Int. J. Sci. Res. Comput. Sci. Eng. Inf. Technol. 2020, 6, 280–293. [CrossRef]

5. Plakida, N. (2010). High-Temperature Cuprate Superconductors Experiment, Theory, and Applications. In Springer Series in Sol-id-State Sciences; Springer: Berlin/Heidelberg, Germany, 2010; ISSN 01711873.

6. Alexandrev, A.S. (2003). Theory of Superconductivity, from Weak to Strong Coupling. Series in Condensed Matter Physics; Institute of Physics Publishing: Philadelphia, PA, USA, 2003.

7. Khare, N. (2003). Handbook of High-Temperature Superconductor; Marcel Dekker: New York, NY, USA, 2003; ISBN 0824708237.

8. Buschow, K. (2003). Magnetic & Superconducting Materials, 2nd ed.; Elsevier: Amsterdam, The Netherlands, 2003.

9. Seidel, P. (2015). Applied Superconductivity;Wiley-VCH: Hoboken, NJ, USA, 2015; Volumes 1 and 2.

10. Drechsler, S.; Mishonov, T. (1998). High-T C Superconductors and Related Materials Material Science, Fundamental Properties, and Some Future Electronic Applications; Springer Science Media: Berlin/Heidenberg, Germany, 1998.

11. Kazufumi, I.; Bangti, J. (2015). Inverse Problems, Tikhonov Theory and Algorithms; Series on Applied Mathematics; World Scientific: Singapore, 2015; Volume 22.
12. Casesnoves, F. (2018). Mathematical Models and Optimization of Erosion and Corrosion. Ph.D. Thesis, Taltech University, Tallinn, Estonia, 14 December 2018; ISSN 25856898.
13. Casesnoves, F. (2019). Die Numerische Reuleaux-Methode Rechnerische und Dynamische Grundlagen mit Anwendungen (Erster Teil); Sciencia Scripts: Chisinau, Moldova, 2019; ISBN 139786200895608/106200895600.
14. Casesnoves, F. (2018). Primary Modelling for Electromagnetic Waves Transmission in Extreme Weather Conditions. Int. J. Innov. Res. Sci. Eng. Technol. 2018, 7, 10294–10300. [CrossRef]
15. Casesnoves, F. (2019). The Numerical Reuleaux Method, a Computational and Dynamical Base with Applications. First Part; Lambert Academic Publishing: Saarland, Germany, 2019; ISBN 103659917478.
16. Darwin, C. (2004). The Origin of Species; Barnes & Noble Classics: New York, NY, USA, 2004.
17. Haupt, R.; Haupt, S. (2004). Practical Genetic Algorithms, 2nd ed.;Wiley: New York, NY, USA, 2004.
18. Abramowitz, M.; Stegun, I.A. (1964). Handbook of Mathematical Functions; U.S. Government Printing Office: Champaign, IL, USA, 1964; Volume 55.
19. Casesnoves, F.; Suzenkov, A. (2017). Mathematical Models in Biotribology with 2D-3D Erosion Integral-Differential Model and Computational-Optimization/Simulation Programming. Int. J. Sci. Res. Comput. Sci. Eng. Inf. Technol. 2017, 2, 329–356; ISSN 24563307.
20. Casesnoves, F.; Antonov, M.; Kulou, Piss. (2016). Mathematical models for erosion and corrosion in power plants. A review of applicable modelling optimization techniques. IEEE Xplore database and will be cross referred in

SCOPUS. In Proceedings of the RUTCON2016 Power Engineering Conference, Riga, Latvia, 13 October 2016.

21. Casesnoves, F. (2018). 2D computational-numerical hardness comparison between Fe-based hardfaces with WC-Co reinforcements for Integral-Differential modelling. Key Eng. Mater. J. 2018, 762, 330–338; ISSN 16629795. [CrossRef]

22. Casesnoves, F.; Surzhenkov, A. (2017). Inverse methods for computational simulations and optimization of erosion models in power plants. In Proceedings of the RUTCON2017 Power Engineering Conference, Riga, Latvia, 5 December 2017. [CrossRef]

23. Casesnoves, F. (2012). Computational Simulations of Vertebral Body for Optimal Instrumentation Design. J. Med. Devices 2012, 6, 021014. [CrossRef]

24. Casesnoves, F. (2007). Large-Scale Matlab Optimization Toolbox (MOT) Computing Methods in Radiotherapy Inverse Treatment Planning. In High Performance Computing Meeting; Nottingham University: Nottingham, UK, 2007.

25. Casesnoves, F. (2007). A Monte-Carlo Optimization method for the movement analysis of pseudo-rigid bodies. In Proceedings of the 10th SIAM Conference in Geometric Design and Computing, San Antonio, TX, USA, 4–8 November 2007. Contributed Talk.

26. Casesnoves, F. (2013). Applied Inverse Methods for Deformable Solid Dynamics/Kinematics in Numerical Reuleaux Method (NRM). Int. J. Numer. Methods Appl. **2013**, 9, 109–131.

27. Casesnoves, F. (2018). Nonlinear comparative optimization for biomaterials wear in artificial implants technology. In Proceedings of the Applied Chemistry and Materials Science RTU2018 Conference Proceedings, Rajasthan Technical University, Kota, India, 17–18 March 2018.

28. Allison, J.; Amako, K.; Apotolakis, J.; Araujo, H.; Dubois, P.A.; Asai, M.; Barrand, G.; Capra, R.; Chauvie, S.; Chytracek, R.; et al. Geant4 Developments and Applications. IEEE Trans. Nucl. Sci. **2006**, 53, 270–278. [CrossRef]

29. Batic, M.; Hoff, G.; Pia, M.G.; Saracco, P.; Weidenspointner, G. (2013). The Geant4 Simulation Toolkit. IEEE Nucl. Sci. Symp. Med. Imaging Conf. Seoul **2013**, 60, 2934–2957.

30. Luenberger, G.D. (2008). Linear and Nonlinear Programming, 4th ed.; Springer: Berlin/Heidelberg, Germany, 2008.

31. Huang, X. (2011). Does the isotope effect of mercury support the BCS theory? arXiv 2011, arXiv:1102.1467.

32. Hummel, R.E. Electronic Properties of Materials; Springer: New York, NY, USA, 2011.

33. Kasap, F.; Capper, P. (2000). (Eds.) Springer Handbook of Electronic and Photonic Materials, 4th ed.; Springer: Berlin/Heidelberg, Germany, 2000. [CrossRef]

34. Kessel,W. (1974). On a General Formula for the Transition Temperature of Superconductors. Naturforschung 1974, 29a, 445–451. [CrossRef]

35. Kulou, P.; Casesnoves, F.; Simson, T.; Tarbe, R. (2017). Prediction of abrasive impact wear of composite hardfacings. Solid State Phenomena. In Proceedings of 26th International Baltic Conference on Materials Engineering; Trans Tech Publications: Bäch, Switzerland, 2017; Volume 267, pp. 201–206. [CrossRef]

36. Luiz, A. (2010). Superconductivity—Theory and Applications; InTechOpen: London, UK, 2010; ISBN 9789533071510.

37. Reynolds, C.A.; Serin, B.; Nesbitt, L.B. The Isotope Effect in Superconductivity. (1951). I. Mercury. Phys. Rev. 1951, 84, 691. [CrossRef]

38. Seri, B.; Reynolds, C.A.; Nesbitt, L.B. (1950). Mass Dependence of the Superconducting Transition Temperature of Mercury. Letters to Editor. Phys. Rev. **1950**, 80, 761. [CrossRef]

39. Todinov, M. (2005). Reliability and Risk Models; Wiley: New York, NY, USA, 2005.

40. Vidyasagar, M. (1993). Nonlinear Systems Analysis, 2nd ed.; Prentice Hall: Hoboken, NJ, USA, 1993.

41. Wesche, R. (2017). Chapter 50. High-Temperature Superconductors. In Springer Handbook of Electronic and Photonic Materials; Springer: Berlin/Heidelberg, Germany, 2017.

42. Parinov, I. (2017). Microstructure and Properties of High-Temperature Superconductors, 2nd ed.; Springer: Berlin/Heidelberg, Germany, 2017.

43. Fossheim, K.; Sudbø, A. (2004). Superconductivity Physics and Applications; Wiley: New York, NY, USA, 2004.

44. Wang, Y. (2013). Fundamental Elements of Applied Superconductivity in Electrical Engineering; Wiley: New York, NY, USA, 2013.

45. European Textbook on Ethics in Research. (2021). European Commission, Directorate-General for Research. Unit L3. Governance and Ethics. European Research Area. Science and Society. EUR 24452 EN. Available online: https://op.europa.eu/en/publicationdetail/-/publication/12567a07-6beb-4998-95cd-8bca103fcf43 (accessed on 28 June 2021).

46. ALLEA. (2017). The European Code of Conduct for Research Integrity, Revised ed.; ALLEA: Bern, Switzerland, 2017.

47. Swedish Research Council. (2017). Good Research Practice; Swedish Research Council: Stockholm, Sweden, 2017; ISBN 978-91-7307-354-7.

3. SCIENTIFIC ETHICS STANDARDS

2D/3D Graphical Optimization Methods were created by Dr Francisco Casesnoves in 3rd November 2016, and Interior Optimization Methods in 2019. 4D Graphical and Interior Optimization Methods were created by Dr Francisco Casesnoves in 2020. MEM and SCMTL were firstly developed by author in 15th March 2021. First paragraph of Chapter IV is taken for an article published later. This GA new software was originally developed by author. This article has previous papers information, whose inclusion is essential to make the contribution understandable. The GA nonlinear optimization software was invented/improved from previous contributions in subroutines modifications, patters, loops, graphics and optimal visualization [2-4]. At Method sections, algorithms for basic Tikhonov Functional were developed from [ch4]. The 4D Interior Optimization method is original from the author (August 2021). This study was carried out, and their contents are done according to the European Union Technology and Science Ethics and International Scientific Community Ethics [45-47].

This research was completely done by the author, the computational-software, calculations, images, mathematical propositions and statements, reference citations, and text is original for the author. When a citation such as [Casesnoves, 'year'] appears, there is not vanity or intention to brag. The reason is to set clearly, at present research times, the intellectual property. When a mathematical statement, proposition or theorem is presented, demonstration is always included if it was not set in previous papers. When any results inconsistency is found after publication, it is clarified in subsequent contributions. The book is exclusively scientific, without any commercial, institutional, academic, religious, religious-similar, non-scientific theories, personal opinions, political ideas, or economical influences. When anything is taken from a source, it is adequately recognized. Ideas and some text expressions/sentences from previous publications were emphasized due to a clarification aim [45-47].

4. AUTHOR'S BIOGRAPHY

Dr Francisco Casesnoves earned the Engineering and Natural Sciences PhD by Talllinn University of Technology (started thesis in 2016, thesis Defence/PhD earned in December 2018, official graduate Diploma 2019). Dr Casesnoves is European Union and Internationally qualified as Doctor in Engineering to supervise PhD Theses, Master Theses, and Bachelor Theses in science and engineering. He works as independent research scientist in computational-engineering/physics.

Dr Casesnoves earned MSc-BSc, Physics/Applied-Mathematics (Public Eastern-Finland-University, MSc Thesis in Radiotherapy Treatment Planning Optimization, which was developed after graduation in a series of Radiation Therapy Optimization-Modelling publications [2007-present]). Dr Casesnoves earned Graduate-with-MPhil, in Medicine and Surgery [1983] (Madrid University Medicine School, MPhil in Radioprotection Low Energies Dosimetry [1985]). He studied always in public-educational institutions, was football player 1972-78 (defender and midfielder) and as Physician, supports healthy life and all sports activities. Casesnoves resigned definitely to his original nationality in 2020 for ideological reasons, democratic-republican ideology, ethical-professional reasons, anti-state monarchy corruption positions, and does not belong to Spain Kingdom anymore. His constant service to the International Scientific Community and Estonian technological progress (2016-present) commenced in 1985 with publications in Medical Physics, with further specialization in optimization methods in 1997 at Finland—at the moment approximately 100 recognized publications with approximately 65 DOI papers. His main branch is Computational-mathematical Nonlinear/Inverse Methods Optimization. Casesnoves best-achievements are the Numerical Reuleaux Method in dynamics and nonlinear-optimization [books 2019-2020],

The series of Radiotherapy Improvements for AAA superposition-convolution model, the Graphical and Interior Optimization Methods [2016-8], the new Computational Dissection-Anatomical Method, [2020], invention of Forensic Robotics [2020-2021], and Molecular Effect Model for High Temperature Superconductors [2020]. Dr Casesnoves PhD thesis is an Estonian scientific service to European Social Fund and several EU Research Projects. Dr Casesnoves scientific service since 2016 to the Free and Independent Republic of Estonia for technological development (and also at Riga technical University, Power Electrical and Electronics Department) is about 40 physics-engineering articles, two books series, and 1 industrial radiotherapy project associated to Europe Union EIT Health Program (Tartu University, 2017).

5. BRIEF INDEX

Publisher: Eliva Press Global Ltd

Email: info@elivapress.com

Made in United States
North Haven, CT
04 November 2026

21772738R00063